TESTS-ANSWERS FOR FCC GENERAL RADIOTELEPHONE OPERATOR LICENSE

by
Warren Weagant

Radio Engineering Division
P.O. Box 2824
San Francisco, CA 94126-2824

FIFTEENTH EDITION - FIRST PRINTING

Library of Congress Cataloging in Publication Data Main Entry Under Title:

TESTS-ANSWERS FOR FCC GENERAL
RADIOTELEPHONE OPERATOR LICENSE

Printed in the United States of America

International Standard Book Number (ISBN): 0-933132-15-8

Contents

Introduction

The purpose of this testing manual is to prepare you to pass the Federal Communication's General Radiotelephone Operator License examinations. This guidebook contains the entire set of questions asked by the FCC, and provides an excellent means of preparation for the federal license exam.

A General Radiotelephone Operator License must be held by any person who adjusts, maintains, or internally repairs a radiotelephone transmitter at any station licensed by the FCC in the aviation, maritime, or international fixed public radio services.

This license must also be held by the operator of certain aviation and maritime land radio stations, compulsory equipped ship radiotelephone stations transmitting with more than 1,500 watts of peak envelope power output, and voluntarily equipped ship stations transmitting with more than 1000 watts peak envelope power. Also, portable transceivers used to communicate with ships and aircraft on marine and aviation frequencies.

The FCC examination for the General Radiotelephone Operator License currently consists of 100 multiple-choice questions. You must pass FCC exam element #1 and #3 to obtain a license. Element #1 contains 24 Rules and Regulations questions. Eighteen (18) questions answered correctly is the passing grade. Element #3 contains 76 technical and operations questions. You must answer 57 questions correctly to pass.

The holder of a General Radiotelephone License must obtain a special FCC endorsement to legally install, maintain and service ship radar equipment. To obtain the radar license endorsement for your General Radiotelephone license, you must pass the FCC exam (FCC Element #8), consisting of 50 multiple-choice questions. Thirty-eight (38) answered correctly is the passing grade. This endorsement is not necessary for most license applicants, only for those people entering the ship-radar and avionics installation and maintenance radio services.

Examinations are no longer given at FCC Field Offices. Beginning in September 1993, the FCC delegated authority to a number of private professional technical organizations (see page 10).

When you are ready to take the FCC License exam you will need to contact one of these organizations to request an examination application. Since each organization has a different schedule of examination dates for each area of the country - you may want to phone, FAX or write to several organizations to determine the best testing date and location that fits your needs. Also, the cost of taking the exam varies; ranging from about $35 to $120.

There is no shortage of text books on electronic theory on the market. Many license applicants attempt to learn all facets of electronics while preparing for the federal exam. However, since the FCC exam covers only certain specific areas of electronic knowledge, you will save hours of study time by using this manual as your learning outline. When you discover any questions in this manual you cannot answer or understand then look up the material in any one of many excellent reference books and courses containing detailed discussion of essential electronic theory.

This testing manual concentrates on specific areas covered on the federal exam. An all-encompassing, general background is necessary for actual station operation and equipment maintenance, but specialized knowledge is required to successfully pass the FCC exam. If the material in this manual is completely learned and understood, you should easily answer questions on the FCC examination. It is important to note that the questions in this guide are the exact words of the questions on the actual FCC exam, although the multiple-choice answers may appear in a different order on the FCC exam.

Finally, you will notice that the format of this manual is concise and comprehensive. Everything included is relevant to the actual FCC exam. Great effort has been made to eliminate all non-related material used by other publishers to "fatten-out" their study guides. This is to point out that it is important to learn all test material in this training manual!

Studying on your own for the FCC examination is no easy task. It requires dedication and a great desire to operate and service sophisticated federally licensed transmitters. The additional money you will earn, the security and the attractive employment opportunities are certainly worth the effort.

Warren Weagant
San Francisco, California

Operator Licenses

A General Radiotelephone Operator License is required for anyone responsible for internal repairs, maintenance, and adjustment of FCC Licensed transmitters in the Aviation, Marine and International Public Fixed radio services.

To be eligible for this license, you must:

1. Be a legal resident - eligible for employment in the United States, including all U.S. Citizens, U.S. Nationals, and citizens of U.S. trust territories.

2. Be able to receive and transmit spoken messages in the English language.

3. Pass a multiple-choice examination covering basic radio law, operating procedures and basic electronic theory.

There are no education, training or experience requirements to obtain an operator license. Knowledge of the telegraphic code is not necessary. Your age is not important. Anyone, regardless of age, can apply for the license. This license is normally valid for the lifetime of the operator.

FCC EXAMINATIONS

Element #1 Examination: Rules and Regulations. This is the preliminary basic requirement for higher class FCC licenses. The test covers provisions of laws, treaties and regulations with which every radio operator should be familiar. Radio operating procedures and practices generally followed or required in communicating by radiotelephone.

From a question pool of 170 possible questions, a test administrator will select 24 questions at random. You must correctly answer 75% - a total of 18 questions before a higher element will be graded.

Element #3 Examination certifies that the examinee possesses the operational and technical qualifications necessary to properly perform the duties required of a person holding a General Radiotelephone Operator License (GROL). To pass, you must answer correctly at least 57 out of 76 questions on this examination.

Each Element 3 examination is administered by a Commercial Operator License Examination Manager (COLEM). The COLEM must construct the examination from this listing using the following algorithm:

- 3A - Operating Procedures - 3 questions
- 3B - Radio Wave Propagation - 3 questions
- 3C - Radio Practice - 5 questions
- 3D - Electrical Principles - 16 questions
- 3E - Circuit Components - 13 questions
- 3F - Practical Circuits - 22 questions
- 3G - Signals and Emissions - 9 questions
- 3H - Antennas and Feed Lines - 5 questions

Element 8 Examination (Radar Endorsement) - Specialized theory and practice applicable to proper installation, servicing, and maintenance of radar equipment. This endorsement is not required for most license applicants - only for those persons working with radar navigation equipment. The FCC exam contains 50 questions. The passing grade is 37 questions answered correctly.

There is no limit to the number of times you may take the FCC exam. Each private testing organization may have a different waiting period between examination testing dates. However, you could simply apply to additional organizations for another testing appointment without delay. This is a much better system than in past years. Before 1993, the FCC offered examinations only twice per year, during specific one week periods.

In September 1993, the FCC approved the testing for all Commercial Radiotelephone Operator License exams by the private sector. It authorized certain national private professional associations and commercial testing organizations to be classified as "Commercial Operator Licensing Examination Managers" - called "COLEM".

When you want to take your FCC examination, simply contact the national headquarters of a COLEM to request an examination appointment. A listing of all national COLEMS is on page 10 in this guidebook. You may simply telephone, FAX or write to any of the COLEMS to request testing applications and forms.

After you receive the application forms in the mail, you will need to fill out the exam application and return it with your payment for the requested amount - usually between $35 and $120.

Caution: You are not required to purchase additional training from these organizations. We have learned that some license applicants are being pressured to purchase expensive training materials at a price of $500 and much more! This is not necessary or required. Remember, all questions that will appear on any FCC examination are contained in this study guide!

When you receive the testing application, you should fill in the city or location where you want to be examined. By return mail, you will be sent a specific testing date, address, procedures, etc. The national organization will generate the actual examination from the same group of questions that appear in this guide. If you are tested for both Element #1 and #3 (recommended), 100 questions will be randomly selected for your examination. Your name and test ID number will be attached to the exam and it will be forwarded to the testing site prior to the scheduled examination date.

When you arrive at the testing location, the local testing proctor may verify your identity by inspecting your driver's license, passport, or other "photo" style identification.

Your FCC exam will not be graded locally. The test will be returned to the national COLEM headquarters. You will be sent a "Proof of Passing Certificate" within 10 days, along with instructions to complete the FCC licensing process. Although the examination process is now contracted to private testing administrators, the actual license is officially issued by the federal government FCC Licensing Division, in Gettysburg, Pennsalvania.

The FCC decision to authorize private professional organizations to help with the examination process has provided more frequent exam opportunities in more accessible locations. You now have the opportunity to take the FCC exam several times per year. Although each COLEM organization may have a required "waiting period" between examination attempts, you may also apply to other COLEMS for additional testing dates. For example, one COLEM may allow you to sit for the examination only once quarterly (in March, May, September and December). If you fail the test, this COLEM requires a 90 day waiting period prior to any scheduled retake.

However, you may request examination appointments from more than one COLEM! If you can afford to submit testing fees to several COLEMS, you could actually attempt the exam more than once in a brief period of time. If your job or career depends upon the license - it might be worth the time, effort and expense.

The FCC has adopted a new style license certificate to replace the old diploma size certificate. Currently, the FCC issues a 2 by 3 inch size license card. This new smaller size is designed to be kept on your person while operating and maintaining equipment in the field. The FCC license is issued for the lifetime of the holder. You are not legally required to obtain additional training, or special continuing education after you get your license, although it is recommended.

Special "Broadcast Radio and TV" authorization: Persons who operate Commercial Radio-TV transmitters may also want to obtain a "Restricted Radiotelephone Operator Permit" from the FCC. This permit is easy to obtain. You may apply for the special limited operating permit simply by requesting FCC Form #753. There is no examination required for this permit, but to be eligible for this permit you must be at least 14 years old; be a legal resident (eligible for employment) in the U.S. and be able to speak and hear, keep a written log in english and be familiar with provisions of applicable treaties, laws and FCC Regulations. Contact the FCC directly for this operator permit.

How to Apply for an FCC License

Examinations are no longer given at FCC Field Offices. Beginning in September 1993, the FCC delegated actual licensee testing administration to a group of private organizations.

If you are planning to take the FCC License exam in the near future, it is advisable to contact one of the following organizations to request an examination appointment.

You may telephone, FAX or write a request letter to one or more of the following organizations to determine the best test date and location that meets your requirements. Also, the cost of taking the exam varies from about $35 to $120.

The first four listings are leading professional associations:

1. The National Association of Radio Telecommunications Engineers, Inc. (NARTE). P.O. Box 678, Medway, MA. 02053. Phone (508) 533-8333. FAX (817) 548-9594. All exam testing is available by appointment on a quarterly basis at test centers located at over 120 universities and colleges throughout the U.S. and at some overseas U.S. military installations. Test applicants should have their application sent to NARTE four weeks prior to examination testing date. Fee: $40.00.

2. National Association of Business and Educational Radio, Inc. (NABER). 1501 Duke Street, Alexandria, VA. 22314. Phone (800) 869-1100. FAX (817) 921-3741. Exams are offered at over 95 test centers nationwide. Fee: From $63.00 to $120.00.

3. International Society of Certified Electronics Technicians (ISCET). 2708 Berry Street., Fort Worth, TX. 76109. Phone (817) 921- 9101. FAX (817) 921-3741. Exams offered in most states and in some foreign countries. Fee: $35.00.

4. Electronic Technicians Association International, Inc. 602 North Jackson Street, Greencastle, IN. 46135. Phone (317) 653-4301. FAX (317) 653-8262. Exams offered throughout the U.S. and at some overseas military installations. Fee: $35.00.

Also, the license exam is conducted by these other private testing service administrators:

Drake Training and Technologies. 8800 Queen Avenue South, Bloomington, MN. 55431. Phone (612) 921-7441. FAX (612) 921-7248. Fee: $60.00.

Sylvan Kee Systems. 9135 Guilford Road, Columbia, MD. 21046. Phone (800) 967-1100. FAX (410) 880-8714. Fee: $50.00 to $75.00.

Sea School. 5905 4th Street North, St. Petersburg, FL. 33703. Phone (800) 237-8663. FAX (813) 522-3155. Fee: $55.00.

Elkins Institute, Inc. P.O. Box 797666, Dallas, TX. 75379. Phone (800) 944-1603. FAX (214) 732-0244. Fee: $75.00.

W5YI Group Inc. P.O. Box 565206, Dallas, TX. 75356. Phone (817) 461-6443 FAX (817) 548-9594.

Study Tips

Test yourself frequently with the questions in this manual. Use your reference books to clear up learning difficulties in the questions missed.

For best results, do not underline or mark the correct answers in this guide. Searching through the wrong answers for the correct statement provides realistic practice for the FCC exam.

Entire books have been written on the subject of how to take multiple-choice exams. Careful study of the exams in this manual should give you ample review of all possible deceptions to the correct answers. Remember, often the exact answer to a question is not one of the possible choices. So, select the nearest correct answer. Also, you may discover the correct answer to be: None of the above, All of the above, Any of the above, Both A and B, etc.

Caution: Spelling errors may appear on FCC exams! The official exam questions in this manual are the exact word by word questions and answers obtained from the FCC. We have included all misspellings and typographical errors in this guide, because they may also appear on the FCC exam you receive. After studying the material in this manual, you should be ready for all questions on the FCC exam.

For best results, learn and understand how to solve math problems and electronic theory questions. Memorize: terminology, rules and regulations, diagrams, tolerances and definitions, etc. When solving math problems, keep your formulas and figuring in a clear logical order. All problems should be saved for suture reference.

When reading the multiple-choice answers in this manual, keep in mind that the actual A, B, C and D answers may appear in a different order on the FCC exam you receive. The COLEM may change the order of answers, but not the actual wording. (Before 1993 there were five possible answers. The exams are now easier - only four multiple-choice selections; the correct answer is grouped with only three distractors.

Please note: This manual contains newly revised updated test questions and answers for passing current FCC exams. You will need to know virtually all information in this guide to pass the FCC license exam.

Question Probability Ratio

You should know that this manual is divided into nine sections of questions. One section for Element #1 material and eight sections for Element #3 questions.

Under ideal conditions, you should study all of the material in this guide to fully prepare for the FCC examination. However, the FCC has specified exactly how many questions may be asked from each section on every exam. Each section contains a different number of possible questions, and the FCC has specified the precise quantity from each section.

Element #1: There are 170 possible questions to study. The actual exam will contain only 24 questions. So, only 14% of the 170 questions will be on your exam. This is a ratio of one out of every seven questions. This is a "Question Probability Ratio" (QPR) of 1:7. Please remember that it is vital that you pass Element #1 - because it is the prerequisite to passing Element #3 exams.

Element #3: The possible questions for this element are divided into eight sections. There are 728 possible questions in element #3. The actual FCC exam for element #3 contains only 76 questions. This means that your exam will contain only about 10% of the total possible questions.

At first glance, it would appear that all questions in the eight sections are of equal importance. But this is not true. Since we know the total number of possible questions in each section, and also know the exact number of questions that the COLEM will select from each section, it is possible to calculate the importance, or "QPR", for each of the 728 questions.

The exclusive "QPR" calculations in this guidebook reveals the relative importance of each Element #3 question.

The first five groups on the next page account for 448 of the 728 possible questions. From the first five groups, a total of 63 questions will be selected on the FCC exam. This is a ratio of 1:7 (one out of every seven questions will be on your test).

IMPORTANCE	SECTION	TOTAL QUESTIONS	QUESTIONS PER EXAM	"QPR"	PERCENT OF TEST
1st	3E	75	13	1:6	17%
2nd	3F	139	22	1:6	29%
3rd	3D	115	16	1:7	21%
4th	3B	22	3	1:7	4%
5th	3G	97	9	1:11	12%
6th	3A	40	3	1:13	4%
7th	3C	97	5	1:19	6.5%
8th	3H	143	5	1:28	6.5%

Sections 3A, 3C and 3H account for 281 of the 728 questions in Element #3 exams. A total of only 13 questions will be selected from these three sections. This is an QPR ratio of 1:22 - an average of only one out of every 22 questions will be on your exam.

Test 3-H is the least important exam section. The actual ratio for 3-H is 1:28 - one out of almost 28 questions will be on your FCC exam. So if your time is limited, it is suggested that you begin your study by learning the material in Element #1, then continue on through the other sections in the above order of importance. (3E, 3F, 3D, 3B, and 3G, followed by on 3A, 3C and 3H).

Mathematical problem solving questions: Most questions that require a math solution are found in section 3D. From a total of 115 possible questions, over 50% require mathematical problem solving. Also, almost 30 more math type questions appear on sections 3C and 3H.

Based on the above calculations, you can expect about 12 math questions to be on your FCC exam, if questions are selected fairly. If your exam is created by computer, without regard to math vs. theory and procedure questions, your test may contain a different quantity.

For example, if only 16 questions may be selected from section 3D, and if approximately 50% of the 115 questions are mathematical - then your exam should contain only about 8 mathematical questions selected from section 3D. Or, if questions are selected totally randomly, it is possible that all or none of the questions could be mathematical.

If you are concerned with math questions, a little extra study time could be set aside for section 3D. Since many math problems look very similar, it is not recommended that you memorize these questions. It is suggested that you learn how to work the formulas and solve the problems. There is plenty of other material to memorize for the exam without trying to memorize math problem answers too.

Before Taking the FCC Exam

Get a full nights sleep. The benefits to be gained by a few extra hours of study are more than offset by the detrimental effects of fatigue.

Don't review or study on the day of the test. You might lose confidence in your ability and not do as well.

Bring the following items with you to the testing sight:

1. Photo identification; Driver's license, Passport, etc.

2. Several pencils and two ball point pen (blue or black ink).

3. Wear comfortable shoes and clothing. Most people spend several hours sitting on a hard uncomfortable chair while taking the examination. And remember, some examiners may not allow you to leave the testing room to go to the rest room during the exam.

4. Electronic Calculator. Only hand-held, battery operated models (not programmable) may be brought to the examination room.

Be sure to ask the examiner how much time you will have to complete taking your test. In past years, there was a minimum of 3 hours time set aside per exam sitting.

Have a good mental attitude. You may miss 25% of the questions in each element and still pass the exam. Also, you should find the actual FCC examination no more difficult than the exams in this guidebook. On the following pages, you will find all FCC exam questions that will appear on your exam!

Taking the FCC Exam

You may be required to fill out several information forms. However, most personal information questions are usually filled out on the original application when you applied to take the examination. Take your time filling out all forms! Any errors on these forms may cause you to flunk the exam. If you have any questions on these forms, do not hesitate to ask the examiner for help.

Answer the exam questions in this order: (1) Answer easy questions first; rules and regulations, terms, etc. (2) Answer easiest theory questions next. (3) Solve easiest math problems. (4) Go back and solve the difficult questions you skipped.

When reading each FCC question, cover the answers with your answer sheet. Attempt to answer the question, THEN read each answer carefully. Remember, the multiple-choice answers may be in a different order than in this manual.

Some of the math questions may NOT be exact figures, so you should select the nearest correct answer. All math problems are included on the following pages.

Answer EVERY question, even if you have to guess! Currently, there are only four muiltiple-choice answers. So, you have a 25% chance of of guessing the correct answer. Before 1993, there were five possible answers for each FCC question - so the the exam is easier now.

After completion of the FCC exam, double check to see if you have marked the answer sheet correctly. Since you will be skipping difficult questions in the first place, be sure to skip the appropriate spaces on the answer sheet.

Spelling errors on the FCC exam may occur. The exam questions in this manual contain the same misspellings, because we want you to be familiar with errors released by the federal government. You may find the same errors on your test. Just because a question has a spelling error, or poor word usage, etc., it may still be correct. You will find all such errors reproduced on purpose in this guide so you will be comfortable selecting answers exactly as they appear on the federal exam. However, some examiners may make the effort to correct certain spelling errors - so be prepared to find either or both on your exam.

All schematic diagrams on the federal exam are contained in this study guide. Actually, the FCC obtained these diagrams from the 14th edition of the author's guidebook.

After Taking the FCC Exam

If you successfully pass the FCC license examination, the COLEM test administration headquarters will mail to you a "Proof of Passing Certificate" (PPC). This form shows that you have passed and receive credit for the examination elements attempted.

You should expect that this certificate should be sent to you within 10 days of taking the exam.

The COLEM private testing service organization will NOT issue the official federal FCC license. It is issued exclusively by the Federal Communications Commission.

After you have received the "Proof of Passing Certificate", you must complete FCC Application Form #756 and mail both forms to:

Federal Communications Commission
1270 Fairfield Road
Gettysburg, PA. 17325-7245

It is recommended that you photocopy both forms prior to sending them to the FCC. If your letter is lost or delayed, a photocopy may be important proof of your passing grade.

Some COLEMS may include a copy of FCC Form #756 with your PPC certificate. If not, contact an FCC Field Office (Page 219) or the main office of the FCC:

Federal Communications Commission
1919 "M" Street, N. W.
Washington, DC 20554

You should NOT send your license application forms to the Washington, DC address. It is only for requesting form #756, or if you have any questions about FCC licensing, etc. The telephone number for the department in charge of licensing is: (202) 632-7240.

If you have any problems or questions about the formalities of taking the FCC exam, obtaining your license certificate, etc., not answered by the Commission or a COLEM, please write to Command Productions and we will try to answer your questions, or route you in the right direction.

FCC Element One Questions

Element 1 (formally Elements 1 and 2): Basic radio law and operating practice with which every maritime radio operator should be familiar. 24 questions concerning provisions of laws, treaties, regulations, and operating procedures and practices generally followed or required in communicating by means of radiotelephone stations. The minimum passing score is 18 questions answered correctly.

The 24 questions asked on your Element 1 examination will be taken from the following group of 170 questions.

Although you may obtain a Marine Radio Operator Permit (MROP) after passing the Element 1 exam, most applicants prefer to continue on to Element 3 and obtain the General Radiotelephone Operator License. You may NOT hold both a GROL and a MROP. Since the GROL is a higher class license - most applicants test for Element 3 immediately after taking the Element 1 exam.

Element 1 is also a partial requirement for any class of Radiotelegrapher's Certificate, any class of Global Maritime Distress and Safety System License and of course the General Radiotelephone License.

For your information, the MROP must be held by any person who operates certain radiotelephone stations aboard voluntarily equipped vessels, some categories of aviation radiotelephone stations, and certain coast stations. A MROP must also be held by the operator of compulsorily equipped radiotelephone stations aboard vessels of more than 300 gross tons, and vessels that carry more than six passengers for hire in the open sea or in any tidewater area of the United States, and certain ships that operate on the Great Lakes.

(1) What is the Global Maritime Distress and Safety System (GMDSS)?

(A) An automated ship-to-shore distress alerting system using satellite and advanced terrestrial communications systems.
(B) An emergency radio service employing analog and manual safety apparatus.
(C) An association of radio officers trained in emergency procedures.
(D) The international organization charged with the safety of ocean-going vessels.

(2) What authority does the Marine Radio Operator Permit confer?

(A) Grants authority to operate commercial broadcast stations and repair associated equipment.
(B) Allows the radio operator to maintain equipment in the Business Radio Service.
(C) Confers authority to operate licensed radio stations in the Aviation, Marine and International Fixed Public Radio Services.
(D) The non-transferable right to install, operate and maintain any type-accepted radio transmitter.

(3) Which of the following persons are ineligible to be issued a commercial radio operator license?

(A) Individuals who are unable to send and receive correctly by telephone spoken messages in English.
(B) Handicapped persons with uncorrected disabilities which affect their ability to perform all duties required of commercial radio operators.
(C) Foreign maritime radio operators unless they are certified by the International Maritime Organization (IMO).
(D) U.S. Military radio operators who are still on active duty.

(4) Who is required to make entries on a required service or maintenance log?

(A) The licensed operator or a person whom he or she designates.
(B) The operator responsible for the station operation or maintenance.
(C) Any commercial radio operator holding at least a Restricted Radiotelephone Operator Permit.
(D) The technician who actually makes the adjustments to the equipment.

(5) What is a requirement of every commercial operator on duty and in charge of a transmitting system?

(A) A copy of the Proof-of-Passing Certificate (PPC) must be on display at the transmitter location.

(B) The original license or a photocopy must be posted or in the operator's personal possession and available for inspection.

(C) The FCC Form 756 certifying the operator's qualifications must be readily available at the transmitting system site.

(D) A copy of the operator's license must be supplied to the radio station's supervisor as evidence of technical qualification.

(6) What is distress traffic?

(A) In radiotelegraphy, SOS sent as a single character; in radiotelephony, the speaking of the word, "Mayday."

(B) Health and welfare messages concerning the immediate protection of property and safety of human life.

(C) Internationally recognized communications relating to emergency situations.

(D) All messages relative to the immediate assistance required by a ship, aircraft or other vehicle in imminent danger.

(7) What is a maritime mobile repeater station?

(A) A fixed land station used to extend the communications range of ship and coast stations.

(B) An automatic on-board radio station which facilitates the transmissions of safety communications aboard ship.

(C) A mobile radio station which links two or more public coast stations.

(D) A one way low-power communications system used in the maneuvering of vessels.

(8) What is an urgency transmission?

(A) A radio distress transmission affecting the security of humans or property.

(B) Health and welfare traffic which impacts the protection of on-board personnel.

(C) A communications alert that important personal messages must be transmitted.

(D) A communications transmission concerning the safety of a ship, aircraft or other vehicle, or of some person on board or within sight.

(9) What is a ship earth station?

(A) A maritime mobile-satellite station located at a coast station.

(B) A mobile satellite station located on board a vessel.

(C) A communications system which provides line-of-sight communications between vessels at sea and coast stations.

(D) An automated ship-to-shore distress alerting system.

(10) What is the internationally recognized urgency signal?

(A) The letters "TTT" transmitted three times by radiotelegraphy.
(B) Three oral repetitions of the word "safety" sent before the call.
(C) The word "PAN" spoken three times before the urgent call.
(D) The pronouncement of the word "Mayday."

(11) What is a safety transmission?

(A) A radiotelephony warning preceded by the words "PAN."
(B) Health and welfare traffic concerning the protection of human life.
(C) A communications transmission which indicates that a station is preparing to transmit an important navigation or weather warning.
(D) A radiotelegraphy alert preceded by the letters "XXX" sent three times.

(12) What is a requirement of all marine transmitting apparatus used aboard United States vessels?

(A) Only equipment that has been type accepted by the FCC for Part 80 operations is authorized.
(B) Equipment must be approved by the U.S. Coast Guard for maritime mobile use.
(C) Certification is required by the International Maritime Organization (IMO).
(D) Programming of all maritime channels must be performed by a licensed Marine Radio Operator.

(13) Where do you submit an application for inspection of a ship radio station?

(A) To a Commercial Operator Licensing Examination Manager (COLE Manager).
(B) To the Federal Communications Commission, Washington, DC 20554.
(C) To the Engineer-in-Charge of the FCC District Office nearest the proposed place of inspection.
(D) To the nearest International Maritime Organization (IMO) review facility.

(14) What are the antenna requirements of a VHF telephony coast, marine utility or ship station?

(A) The shore or on-board antenna must be vertically polarized.
(B) The antenna array must be type accepted for 30-200 MHz operation

by the FCC.

(C) The horizontally polarized antenna must be positioned so as not to cause excessive interference to other stations.

(D) The antenna must be capable of being energized by an output in excess of 100 watts.

(15) What regulations govern the use and operation of FCC-licensed ship stations operating in international waters?

(A) The regulations of the International Maritime Organization (IMO) and Radio Officers Union.

(B) Part 80 of the FCC Rules plus the international Radio Regulations and agreements to which the United States is a party.

(C) The Maritime Mobile Directives of the International Telecommunication Union.

(D) Those of the FCC's Aviation and Marine Branch, PRB, Washington, DC 20554.

(16) Which of the following transmissions are not authorized in the Maritime Service?

(A) Communications from vessels in dry dock undergoing repairs.

(B) Message handling on behalf of third parties for which a charge is rendered.

(C) Needless or superfluous radiocommunications.

(D) Transmissions to test the operating performance of on-board station equipment.

(17) What are the highest priority communications from ships at sea?

(A) All critical message traffic authorized by the ship's master.

(B) Navigation and meteorological warnings.

(C) Distress calls, and communications preceded by the international urgency and safety signals.

(D) Authorized government communications for which priority right has been claimed.

(18) What is the best way for a radio operator to minimize or prevent interference to other stations?

(A) By using an omni-directional antenna pointed away from other stations.

(B) Reducing power to a level that will not affect other on-frequency communications.

(C) By changing frequency when notified that a radiocommunication causes interference.

(D) Determine that a frequency is not in use by monitoring the frequency before transmitting.

(19) Under what circumstances may a ship or aircraft station interfere with a public coast station?

(A) Under no circumstances during on-going radiocommunications.
(B) During periods of government priority traffic handling.
(C) When it is necessary to transmit a message concerning the safety of navigation or important meteorological warnings.
(D) In cases of distress.

(20) Who determines when a ship station may transmit routine traffic destined for a coast or Government station in the maritime mobile service?

(A) Shipboard radio officers may transmit traffic when it will not interfere with on-going radiocommunications.
(B) The order and time of transmission and permissible type of message traffic is decided by the licensed on-duty operator.
(C) Ship stations must comply with instructions given by the coast or Government station.
(D) The precedence of conventional radiocommunications is determined by FCC and international regulation.

(21) Who is responsible for payment of all charges accruing to other facilities for the handling or forwarding of messages?

(A) The licensee of the ship station transmitting the messages.
(B) The third party for whom the message traffic was originated.
(C) The master of the ship jointly with the station licensee.
(D) The licensed commercial radio operator transmitting the radiocommunication.

(22) Ordinarily, how often would a station using a telephony emission identify?

(A) At least every 10 minutes.
(B) At 15 minute intervals unless public correspondence is in progress.
(C) At the beginning and end of each communication and at 15 minute intervals.
(D) At 20 minute intervals.

(23) When does a maritime radar transmitter identify its station?

(A) By radiotelegraphy at the onset and termination of operation.
(B) At 20 minute intervals using an automatic transmitter identification system.
(C) Radar transmitters must not transmit station identification.
(D) By a transmitter identification label (TIL) secured to the

transmitter.

(24) What is the general obligation of a coast or marine-utility station?

(A) To accept and dispatch messages without charge which are necessary for the business and operational needs of ships.
(B) To acknowledge and receive all calls directed to it by ship or aircraft stations.
(C) To transmit lists of call signs of all fixed and mobile stations for which they have traffic.
(D) To broadcast warnings and other information for the general benefit of all mariners.

(25) How does a coast station notify a ship that it has a message for the ship?

(A) By making a directed transmission on 2182 kHz or 156.800 MHz.
(B) The coast station changes to the vessel's known working frequency.
(C) By establishing communications using the eight digit maritime mobile service identification.
(D) The coast station may transmit at intervals lists of call signs in alphabetical order for which they have traffic.

(26) Under what circumstances may a coast station using telephony transmit a general call to a group of vessels?

(A) Under no circumstances.
(B) When announcing or preceding the transmission of distress, urgency, safety or other important messages.
(C) When the vessels are located in international waters beyond 12 miles.
(D) When identical traffic is destined for multiple mobile stations within range.

(27) Who has ultimate control of service at a ship's radio station?

(A) The master of the ship.
(B) A holder of a First Class Radiotelegraph Certificate with a six months service endorsement.
(C) The Radio Officer-in-Charge authorized by the captain of the vessel.
(D) An appointed licensed radio operator who agrees to comply with all Radio Regulations in force.

(28) What is the power limitation of associated ship stations operating under the authority of a ship station license?

(A) The power level authorized to the parent ship station.
(B) Associated vessels are prohibited from operating under the authority granted to another station licensee.
(C) The minimum power necessary to complete the radiocommunications.
(D) Power is limited to one watt.

(29) How is an associated vessel operating under the authority of another ship station license identified?

(A) All vessels are required to have a unique call sign issued by the Federal Communications Commission.
(B) With any station call sign self-assigned by the operator of the associated vessel.
(C) By the call sign of the station with which it is connected and an appropriate unit designator.
(D) Client vessels use the call sign of their parent plus the appropriate ITU regional indicator.

(30) On what frequency should a ship station normally call a coast station when using a radiotelephony emission?

(A) On a vacant radio channel determined by the licensed radio officer.
(B) Calls should be initiated on the appropriate ship-to-shore working frequency of the coast station.
(C) On any calling frequency internationally approved for use within ITU Region 2.
(D) On 2182 kHz or 156.800 MHz at any time.

(31) On what frequency would a vessel normally call another ship station when using a radiotelephony emission?

(A) Only on 2182 kHz in ITU Region 2.
(B) On the appropriate calling channel of the ship station at 15 minutes past the hour.
(C) On 2182 kHz or 156.800 MHz unless the station knows the called vessel maintains a simultaneous watch on another intership working frequency.
(D) On the vessel's unique working radio-channel assigned by the Federal Communications Commission.

(32) What is required of a ship station which has established initial contact with another station on 2182 kHz or 156.800 MHz?

(A) The stations must check the radio channel for distress, urgency and safety calls at least once every ten minutes.
(B) The stations must change to an authorized working frequency for the transmission of messages.

(C) Radiated power must be minimized so as not to interfere with other stations needing to use the channel.

(D) To expedite safety communications, the vessels must observe radio silence for two out of every fifteen minutes.

(33) What type of communications may be exchanged by radioprinter between authorized private coast stations and ships of less than 1600 gross tons?

(A) Public correspondence service may be provided on voyages of more than 24 hours.

(B) All communications providing they do not exceed 3 minutes after the stations have established contact.

(C) Only those communications which concern the business and operational needs of vessels.

(D) There are no restrictions.

(34) What are the service requirements of all ship stations?

(A) Each ship station must receive and acknowledge all communications with any station in the maritime mobile service.

(B) Public correspondence services must be offered for any person during the hours the radio operator is normally on duty.

(C) All Ship stations must maintain watch on 500 kHz, 2182 kHz and 156.800 MHz.

(D) Reserve antennas, emergency power sources and alternate communications installations must be available.

(35) When may the operator of a ship radio station allow an unlicensed person to speak over the transmitter?

(A) At no time. Only commercially licensed radio operators may modulate the transmitting apparatus.

(B) When the station power does not exceed 200 watts peak envelope power.

(C) When under the supervision of the licensed operator.

(D) During the hours that the radio officer is normally off duty.

(36) What are the radio operator requirements of a cargo ship equipped with a 1000 watt peak-envelope-power radiotelephone station?

(A) The operator must hold a General Radiotelephone Operator License or higher class license.

(B) The operator must hold a Restricted Radiotelephone Operator Permit or higher class license.

(C) The operator must hold a Marine Radio Operator Permit or higher class license.

(D) The operator must hold a GMDSS Radio Maintainer's License.

(37) What are the radio operator requirements of a small passenger ship carrying more than six passengers equipped with a 1000 watt carrier power radiotelephone station?

(A) The operator must hold a General Radiotelephone Operator or higher class license.
(B) The operator must hold a Marine Radio Operator Permit or higher class license.
(C) The operator must hold a Restricted Radiotelephone Operator Permit or higher class license.
(D) The operator must hold a GMDSS Radio Operator's License.

(38) Which commercial radio operator license is required to operate a fixed tuned ship radar station with external controls?

(A) A radio operator certificate containing a Ship Radar Endorsement.
(B) A Marine Radio Operator Permit or higher.
(C) Either a First or Second Class Radiotelegraph certificate or a General Radiotelephone Operator License.
(D) No radio operator authorization is required.

(39) Which commercial radio operator license is required to install a VHF transmitter in a voluntarily equipped ship station?

(A) A Marine Radio Operator Permit or higher class of license.
(B) None, if installed by, or under the supervision of, the licensee of the ship station and no modifications are made to any circuits.
(C) A Restricted Radiotelephone Operator Permit or higher class of license.
(D) A General Radiotelephone Operator License.

(40) What transmitting equipment is authorized for use by a station in the maritime services?

(A) Transmitters that have been certified by the manufacturer for maritime use.
(B) Unless specifically excepted, only transmitters type accepted by the Federal Communications Commission for Part 80 operations.
(C) Equipment that has been inspected and approved by the U.S. Coast Guard.
(D) Transceivers and transmitters that meet all ITU specifications for use in maritime mobile service.

(41) What is the Communication Act's definition of a "passenger ship"?

(A) Any ship which is used primarily in commerce for transporting persons to and from harbors or ports.

(B) A vessel that carries or is licensed or certificated to carry more than 12 passengers.
(C) Any ship transporting more than six passengers for hire.
(D) A vessel of any nation that has been inspected and approved as a passenger carrying vessel.

(42) What is a distress communication?

(A) An internationally recognized communication indicating that the sender is threatened by grave and imminent danger and requests immediate assistance.
(B) Communications indicating that the calling station has a very urgent message concerning safety.
(C) Radiocommunications which, if delayed, will adversely affect the safety of life or property.
(D) An official radiocommunications notification of approaching navigational or meteorological hazards.

(43) Who may be granted a ship station license in the maritime service?

(A) Anyone, including foreign governments.
(B) Only FCC licensed operators holding a First or Second Class Radiotelegraph Operator's Certificate or the General Radiotelephone Operator License.
(C) Vessels that have been inspected and approved by the U.S. Coast Guard and Federal Communications Commission.
(D) The owner or operator of a vessel, or their subsidiaries.

(44) Who is responsible for the proper maintenance of station logs?

(A) The station licensee and the radio operator in charge of the station.
(B) The station licensee.
(C) The commercially licensed radio operator in charge of the station.
(D) The ship's master and the station licensee.

(45) How long should station logs be retained when there are entries relating to distress or disaster situations?

(A) Until authorized by the Commission in writing to destroy them.
(B) Indefinitely, or until destruction is specifically authorized by the U.S. Coast Guard.
(C) For a period of three years from the date of entry unless notified by the FCC.
(D) For a period of one year from the date of entry.

(46) Where must ship station logs be kept during a voyage?

(A) At the principal radiotelephone operating position.
(B) They must be secured in the vessel's strongbox for safekeeping.
(C) In the personal custody of the licensed commercial radio operator.
(D) All logs are turned over to the ship's master when the radio operator goes off duty.

(47) What is the antenna requirement of a radiotelephone installation aboard a passenger vessel?

(A) The antenna must be located a minimum of 15 meters from the radiotelegraph antenna.
(B) An emergency reserve antenna system must be provided for communications on 156.8 MHz.
(C) The antenna must be vertically polarized and as non-directional and efficient as is practicable for the transmission and reception of ground waves over seawater.
(D) All antennas must be tested and the operational results logged at least once during each voyage.

(48) Where must the principal radiotelephone operating position be installed in a ship station?

(A) At the principal radio operating position of the vessel.
(B) In the room or an adjoining room from which the ship is normally steered while at sea.
(C) In the chart room, master's quarters or wheel house.
(D) At the level of the main wheel house or at least one deck above the ship's main deck.

(49) What are the technical requirements of a VHF antenna system aboard a vessel?

(A) The antenna must provide an amplification factor of at least 2.1 dbi.
(B) The antenna must be vertically polarized and non-directional.
(C) The antenna must be capable of radiating a signal a minimum of 150 nautical miles on 156.8 MHz.
(D) The antenna must be constructed of corrosion-proof aluminum and capable of proper operation during an emergency.

(50) How often must the radiotelephone installation aboard a small passenger boat be inspected?

(A) Equipment inspections are required at least once every 12 months.
(B) When the vessel is first placed in service and every 2 years thereafter.
(C) At least once every five years.

(D) A minimum of every 3 years, and when the ship is within 75 statute miles of an FCC field office.

(51) How far from land may a small passenger vessel operate when equipped only with a VHF radiotelephone installation?

(A) No more than 20 nautical miles from the nearest land if within the range of a VHF public coast or U.S. Coast Guard station.
(B) No more than 100 nautical miles from the nearest land.
(C) No more than 20 nautical miles unless equipped with a reserve power supply.
(D) The vessel must remain within the communications range of the nearest coast station at all times.

(52) What is the minimum transmitter power level required by the FCC for a medium frequency transmitter aboard a compulsorily fitted vessel?

(A) At least 100 watts single side band suppressed carrier power.
(B) At least 60 watts PEP.
(C) The power predictably needed to communicate with the nearest public coast station operating on 2182 kHz.
(D) At least 25 watts delivered into 50 ohms effective resistance when operated with a primary voltage of 13.6 volts DC.

(53) What is a Class "A" EPIRB?

(A) An alerting device notifying mariners of imminent danger.
(B) A satellite-based maritime distress and safety alerting system.
(C) An automatic, battery-operated emergency position indicating radiobeacon that floats free of a sinking ship.
(D) A high efficiency audio amplifier.

(54) What are the radio watch requirements of a voluntary ship?

(A) While licensees are not required to operate the ship radio station, general purpose watches must be maintained if they do.
(B) Radio watches must be maintained on 500 kHz, 2182 kHz and 156.800 MHz, but no station logs are required.
(C) Radio watches are optional but logs must be maintained of all medium, high frequency and VHF radio operation.
(D) Radio watches must be maintained on the 156-158 MHz, 1600-4000 KHz and 4000-23000 kHz bands.

(55) What is the Automated Mutual-Assistance Vessel Rescue System?

(A) A voluntary organization of mariners who maintain radio watch on 500 kHz, 2182 kHz and 156.800 MHz.

(B) An international system operated by the Coast Guard providing coordination of search and rescue efforts.
(C) A coordinated radio direction finding effort between the Federal Communications Commission and U.S. Coast Guard to assist ships in distress.
(D) A satellite-based distress and safety alerting program operated by the U.S. Coast Guard.

(56) What is a bridge-to-bridge station?

(A) An internal communications system linking the wheel house with the ship's primary radio operating position and other integral ship control points.
(B) A inland waterways and coastal radio station serving ship stations operating within the United States.
(C) A portable ship station necessary to eliminate frequent application to operate a ship station on board different vessels.
(D) A VHF radio station located on a ship's navigational bridge or main control station that is used only for navigational communications.

(57) Which of the following statements is true as to ships subject to the Safety Convention?

(A) A cargo ship participates in international commerce by transporting goods between harbors.
(B) Passenger ships carry six or more passengers for hire as opposed to transporting merchandise.
(C) A cargo ship is any ship that is not licensed or certificated to carry more than 12 passengers.
(D) Cargo ships are FCC inspected on an annual basis while passenger ships undergo U.S. Coast Guard inspections every six months.

(58) What is a "passenger carrying vessel" when used in reference to the Great Lakes Radio Agreement?

(A) A vessel that is licensed or certificated to carry more than twelve passengers.
(B) Any ship carrying more than six passengers for hire.
(C) Any ship, the principal purpose of which is to ferry persons on the Great Lakes and other inland waterways.
(D) A ship which is used primarily for transporting persons and goods to and from domestic harbors or ports.

(59) How do the FCC's Rules define a power-driven vessel?

(A) A ship that is not manually propelled or under sail.
(B) Any ship propelled by machinery.

(C) A watercraft containing a motor with a power rating of at least 3 HP.

(D) A vessel moved by mechanical equipment at a rate of 5 knots or more.

(60) How do the rules define "navigational communications"?

(A) Safety communications pertaining to the maneuvering or directing of vessels movements.

(B) Important communications concerning the routing of vessels during periods of meteorological crisis.

(C) Telecommunications pertaining to the guidance of maritime vessels in hazardous waters.

(D) Radio signals consisting of weather, sea conditions, notices to mariners and potential dangers.

(61) What traffic management service is operated by the U.S. Coast Guard in certain designated water areas to prevent ship collisions, groundings and environmental harm?

(A) Water safety management bureau (WSMB) .

(B) Vessel traffic service (VTS).

(C) Ship movement and safety agency (SMSA).

(D) Interdepartmental harbor and port patrol (IHPP).

(62) What action must be taken by the owner or operator of a vessel who changes its name?

(A) A Request for Ship License Modification (RSLM) must be submitted to the FCC's licensing facility.

(B) The Engineer-in-Charge of the nearest FCC field office must be informed.

(C) The Federal Communications Commission in Gettysburg, PA, must be notified in writing.

(D) Written confirmation must be obtained from the U.S. Coast Guard.

(63) When may a shipboard radio operator make a transmission in the maritime services not addressed to a particular station or stations?

(A) General CQ calls may only be made when the operator is off duty and another operator is on watch.

(B) Only during the transmission of distress, urgency or safety signals or messages, or to test the station.

(C) Only when specifically authorized by the master of the ship.

(D) When the radio officer is more than 12 miles from shore and the nearest ship or coast station is unknown.

(64) What is the order of priority of radiotelephone communications in the maritime services?

(A) Distress calls and signals, followed by communications preceded by urgency and safety signals.
(B) Alarm, radio-direction finding, and health and welfare communications.
(C) Navigation hazards, meteorological warnings, priority traffic
(D) Government precedence, messages concerning safety of life and protection of property and traffic concerning grave and imminent danger.

(65) What should a station operator do before making a transmission?

(A) Transmit a general notification that the operator wishes to utilize the channel.
(B) Except for the transmission of distress calls, determine that the frequency is not in use by monitoring the frequency before transmitting.
(C) Check transmitting equipment to be certain it is properly calibrated.
(D) Ask if the frequency is in use.

(66) What is the proper procedure for testing a radiotelephone installation?

(A) Transmit the station's call sign, followed by the word "test" on the radio channel being used for the test.
(B) A dummy antenna must be used to insure the test will not interfere with ongoing communications.
(C) Permission for the voice test must be requested and received from the nearest public coast station.
(D) Short tests must be confined to a single working frequency and must never be conduct

(67) What is the minimum radio operator requirement for ships subject to the Great Lakes Radio Agreement?

(A) Third Class Radiotelegraph Operator's Certificate.
(B) General Radiotelephone Operator License.
(C) Marine Radio Operator Permit.
(D) Restricted Radiotelephone Operator Permit.

(68) What FCC authorization is required to operate a VHF transmitter on board a vessel voluntarily equipped with radio and sailing on a domestic voyage?

(A) No radio operator license or permit is required.
(B) Marine Radio Operator Permit.
(C) Restricted Radiotelephone Operator Permit.

(D) General Radiotelephone Operator License.

(69) On what frequencies does the Communications Act require radio watches by compulsory radiotelephone stations?

(A) Watches are required on 500 kHz and 2182 kHz.
(B) Continuous watch is required on 2182 kHz only.
(C) On all frequencies between 405-535 kHz, 1605-3500 kHz and 156-162 MHz.
(D) Watches are required on 2182 kHz and 156.800 MHz.

(70) What is the purpose of the international radiotelephone alarm signal?

(A) To notify nearby ships of the loss of a person or persons overboard.
(B) To call attention to the upcoming transmission of an important meteorological warning.
(C) To alert radio officers monitoring watch frequencies of a forthcoming distress, urgency or safety message.
(D) To actuate automatic devices giving an aural alarm to attract the attention of the operator where there is no listening watch on the distress frequency.

(71) What is the proper procedure for making a correction in the station log?

(A) The ship's master must be notified, approve and initial all changes to the station log.
(B) The mistake may be erased and the correction made and initialled only by the radio operator making the original error.
(C) The original person making the entry must strike out the error, initial the correction and indicate the date of correction.
(D) Rewrite the new entry in its entirety directly below the incorrect notation and initial the change.

(72) What authorization is required to operate a 350 watt PEP maritime voice station on frequencies below 30 MHz aboard a small non-commercial pleasure vessel?

(A) Third Class Radiotelegraph Operator's Certificate.
(B) General Radiotelephone Operator License.
(C) Restricted Radiotelephone Operator Permit.
(D) Marine Radio Operator Permit.

(73) What is selective calling?

(A) A coded transmission directed to a particular ship station.
(B) A radiotelephony communication directed at a particular ship

station.

(C) An electronic device which uses a discriminator circuit to filter out unwanted signals.

(D) A telegraphy transmission directed only to another specific radiotelegraph station.

(74) In the International Phonetic Alphabet, the letters D, N, and O are represented by the words:

(A) Delta, November, Oscar.
(B) Denmark, Neptune, Oscar.
(C) December, Nebraska, Olive.
(D) Delta, Neptune, Olive.

(75) When is it legal to transmit high power on channel 13?

(A) Failure of vessel being called to respond.
(B) In a blind situation such as rounding a bend in a river.
(C) During an emergency.
(D) All of the above.

(76) What must be in operation when no operator is standing watch on a compulsory radio equipped vessel while out at sea?

(A) An auto alarm.
(B) Indicating Radio Beacon signals.
(C) Distress-Alert signal device.
(D) Radiotelegraph transceiver set to 2182 kHz.

(77) When may a bridge-to-bridge transmission be more than 1 watt?

(A) When broadcasting a distress message.
(B) When rounding a bend in a river or traveling in a blind spot.
(C) When calling the Coast Guard.
(D) Both A and B above.

(78) When are EPIRB batteries changed?

(A) After emergency use; after battery life expires.
(B) After emergency use; as per manufacturers instructions marked on outside of transmitter with month and year replacement date.
(C) After emergency use; every 12 months when not used.
(D) Whenever voltage drops to less than 50% of full charge.

(79) The radiotelephone distress message consists of:

(A) MAYDAY spoken three times, call sign and name of vessel in distress.
(B) Particulars of its position, latitude and longitude, and other information which might facilitate rescue, such as length, color and type of vessel, number of persons on board.
(C) Nature of distress and kind of assistance desired.
(D) All of the above.

(80) If a ship sinks, what device is designed to float free of the mother ship, is turned on automatically and transmits a distress signal?

(A) EPIRB on 121.5 MHz/243 MHz or 406.025 MHz.
(B) EPIRB on 2182 kHz and 405.025 kHz.
(C) Bridge-to-bridge transmitter on 2182 kHz.
(D) Auto alarm keyer on any frequency.

(81) International laws and regulations require a silent period on 2182 kHz:

(A) For three minutes immediately after the hour.
(B) For three minutes immediately after the half-hour.
(C) For the first minute of every quarter-hour.
(D) Both A and B above.

(82) How should the 2182 kHz auto-alarm be tested?

(A) On a different frequency into antenna.
(B) On a different frequency into dummy load.
(C) On 2182 KHz into antenna.
(D) Only under U.S. Coast Guard authorization.

(83) What is the average range of VHF marine transmissions?

(A) 150 miles.
(B) 50 miles.
(C) 20 miles.
(D) 10 miles.

(84) A ship station using VHF bridge-to-bridge Channel 13:

(A) May be identified by call sign and country of origin.
(B) Must be identified by call sign and name of vessel.
(C) May be identified by the name of the ship in lieu of call sign.
(D) Does not need to identify itself within 100 miles from shore.

(85) When using a SSB station on 2182 kHz or VHF-FM on Channel 16:

(A) Preliminary call must not exceed 30 seconds.
(B) If contact is not made, you must wait at least 2 minutes before repeating the call.
(C) Once contact is established you must switch to a working frequency.
(D) All of the above.

(86) By international agreement which ships must carry radio equipment for the safety of life at sea?

(A) Cargo ships of more than 300 gross tons and vessels carrying more than 12 passengers.
(B) All ships traveling more than 100 miles out to sea.
(C) Cargo ships of more than 100 gross tons and passenger vessels on international deep-sea voyages.
(D) All cargo ships of more than 100 gross tons.

(87) What is the most important practice that a radio operator must learn?

(A) Monitor the channel before transmitting.
(B) Operate with lowest power necessary.
(C) Test a radiotelephone transmitter daily.
(D) Always listen to 121.5 MHz.

(88) Portable ship radio transceivers operated as associated ship units:

(A) Must be operated on the safety and calling frequency 156.8 MHz (Channel 16) or a VHF intership frequency.
(B) May not be used from shore without a separate license.
(C) Must only communicate with the ship station with which it is associated or with associated portable ship units.
(D) All of the above.

(89) Which is a radiotelephony calling and distress frequency?

(A) 500 kHz.
(B) 2182 kHz.
(C) 156.3 MHz.
(D) 3113 kHz.

(90) What is the priority of communications?

(A) Distress, urgency, safety and radio direction finding.
(B) Safety, distress, urgency and radio direction finding.
(C) Distress, safety, radio direction finding, search and rescue.
(D) Radio direction finding, distress and safety.

(91) Cargo ships of 300 to 1600 gross tons should be able to transmit a minimum range of:

(A) 75 miles.
(B) 150 miles.
(C) 200 miles.
(D) 300 miles.

(92) Radiotelephone stations required to keep logs of their transmissions must include:

(A) Station, date and time.
(B) Name of operator on duty.
(C) Station call signs with which communication took place.
(D) All of the above.

(93) Each cargo ship of the United States which is equipped with a radiotelephone station for compliance with Part II of Title III of the Communications Act shall while being navigated outside of a harbor or port keep a continuous and efficient watch on:

(A) 2182 kHz.
(B) 156.8 MHz.
(C) Both A and B.
(D) Monitor all frequencies within the 2000 kHz to 27500 kHz band used for communications.

(94) What call should you transmit on channel 16 if your ship is sinking?

(A) SOS three times.
(B) MAYDAY three times.
(C) PAN three times.
(D) URGENCY three times.

(95) Under normal circumstances, what do you do if the transmitter aboard your ship is operating off-frequency, overmodulating or distorting?

(A) Reduce to low power.
(B) Stop transmitting.
(C) Reduce audio volume level.
(D) Make a notation in station operating log.

(96) The urgency signal has lower priority than:

(A) Direction finding.
(B) Distress.

(C) Safety.
(D) Security.

(97) The primary purpose of bridge-to-bridge communications is:

(A) Search and rescue emergency calls only.
(B) All short range transmission aboard ship.
(C) Transmission of Captain's orders from the bridge.
(D) Navigational communications.

(98) What is the international VHF digital selective calling channel?

(A) 2182 kHz.
(B) 156.35 MHz.
(C) 156.525 MHz.
(D) 500 kHz.

(99) When your transmission is ended and you expect no response, say:

(A) BREAK.
(B) OVER.
(C) ROGER.
(D) CLEAR.

(100) When attempting to contact other vessels on Channel 16:

(A) Limit calling to 30 seconds.
(B) If no answer is received, wait 2 minutes before calling vessel again.
(C) Channel 16 is used for emergency calls only.
(D) Both A and B.

(101) When a message has been received and will be complied with, say:

(A) MAYDAY.
(B) OVER.
(C) ROGER.
(D) WILCO.

(102) The FCC may suspend an operator license upon proof that the operator:

(A) Has assisted another to obtain a license by fraudulent means
(B) Has willfully damaged transmitter equipment.
(C) Has transmitted obscene language.
(D) Any of the above.

(103) What channel must compulsorily equipped vessels monitor at all times in the open sea?

(A) Channel 8, 156.4 MHz.
(B) Channel 16, 156.8 MHz.
(C) Channel 22A, 157.1 MHz.
(D) Channel 6, 156.3 MHz.

(104) When testing is conducted on 2182 kHz or 156.8 MHz testing should not continue for more than ___________ in any 5 minute period.

(A) 10 seconds.
(B) 1 minute.
(C) 2 minutes.
(D) None of the above.

(105) Which VHF channel is used only for digital selective calling?

(A) Channel 70.
(B) Channel 16
(C) Channel 22A.
(D) Channel 6.

(106) VHF ship station transmitters must have the capability of reducing carrier power to:

(A) 1 watt.
(B) 10 watts.
(C) 25 watts.
(D) 50 watts.

(107) The system of substituting words for corresponding letters is called:

(A) International code system.
(B) Phonetic system.
(C) Mnemonic system.
(D) 10 codes.

(108) How long should station logs be retained when there are no entries relating to distress or disaster situations?

(A) For a period of three years from the date of entry unless notified by the FCC.
(B) Until authorized by the Commission in writing to destroy them.
(C) Indefinitely, or until destruction is specifically authorized by the U.S. Coast Guard.
(D) For a period of one year from the date of entry.

(109) The auto alarm device for generating signals shall be:

(A) Tested monthly using a dummy load.
(B) Tested every three months using a dummy load.
(C) Tested weekly using an dummy load.
(D) None of the above.

(110) Licensed radiotelephone operators are not required on board ships for:

(A) Voluntarily equipped ship stations on domestic voyages operating on VHF channels.
(B) Ship radar, provided the equipment is non-tunable, pulse type magnetron and can be operated by means of exclusively external controls.
(C) Installation of a VHF transmitter in a ship station where the work is performed by or under the immediate supervision of the licensee of the ship station.
(D) Any of the above.

(111) Under what license are hand-held transceivers covered when used on board a ship at sea?

(A) The ship station license.
(B) Under the authority of the licensed operator.
(C) Walkie-talkie radios are illegal to use at sea.
(D) No license is needed.

(112) What should an operator do to prevent interference?

(A) Turn off transmitter when not in use.
(B) Monitor channel before transmitting.
(C) Transmissions should be as brief as possible.
(D) Both B and C.

(113) Identify a ship station's radiotelephone transmissions by:

(A) Country of registration.
(B) Call sign.
(C) Name of the vessel.
(D) Both B and C.

(114) Maritime emergency radios should be tested:

(A) Before each voyage.
(B) Weekly while the ship is at sea.
(C) Every 24 hours.

(D) Both A and B.

(115) The URGENCY signal concerning the safety of a ship, aircraft or person shall be sent only on the authority of:

(A) Master of ship.
(B) Person responsible for mobile station.
(C) Either A or B above.
(D) An FCC licensed operator.

(116) Survival craft emergency transmitter tests may NOT be made:

(A) For more than 10 seconds.
(B) Without using station call sign, followed by the word "test."
(C) Within 5 minutes of a previous test.
(D) All of the above.

(117) International laws and regulations require a silent period on 2182 kHz:

(A) For three minutes immediately after the hour.
(B) For three minutes immediately after the half-hour.
(C) For the first minute of every quarter-hour.
(D) Both A and B above.

(118) How should the 2182 kHz auto alarm be tested?

(A) On a different frequency into antenna.
(B) On a different frequency into dummy load.
(C) On 2182 kHz into dummy load.
(D) On 2182 kHz into antenna.

(119) Each cargo ship of the United States which is equipped with a radiotelephone station for compliance with the Safety Convention shall, while at sea:

(A) Not transmit on 2182 kHz during emergency conditions.
(B) Keep the radiotelephone transmitter operating at full 100% carrier power for maximum reception on 2182 KHz.
(C) Reduce peak envelope power on 156.8 MHz during emergencies.
(D) Keep continuous watch on 2182 kHz using a watch receiver having a loudspeaker and auto alarm distress frequency watch receiver.

(120) What is the procedure for testing a 2182 kHz ship radiotelephone

transmitter with full carrier power while out at sea?

(A) Reduce to low power, then transmit test tone.
(B) Switch transmitter to another frequency before testing.
(C) Simply say: "This is (call letters) testing." If all meters indicate normal values, it is assumed transmitter is operating properly.
(D) It is not permitted to test on the air.

(121) If your transmitter is producing spurious harmonics or is operating at a deviation from the technical requirements of the station authorization:

(A) Continue operating until returning to port.
(B) Repair problem within 24 hours.
(C) Cease transmission.
(D) Reduce power immediately.

(122) As an alternative to keeping watch on a working frequency in the band 1600-4000 kHz, an operator must tune station receiver to monitor 2182 kHz:

(A) At all times.
(B) During distress calls only.
(C) During daytime hours of service.
(D) During the silence periods each hour.

(123) An operator or maintainer must hold a General Radiotelephone Operator License to:

(A) Adjust or repair FCC licensed transmitters in the aviation, maritime and international fixed public radio services.
(B) Operate voluntarily equipped ship maritime mobile or aircraft transmitters with more than 1,000 watts of peak envelope power.
(C) Operate radiotelephone equipment with more than 1,500 watts of peak envelope power on cargo ships over 300 gross tons.
(D) All of the above

(124) What is the radiotelephony calling and distress frequency?

(A) 500 kHz.
(B) 500R122JA.
(C) 2182 kHz.
(D) 2182R2647.

(125) If a ship radio transmitter signal becomes distorted:

(A) Cease operations.
(B) Reduce transmitter power.

(C) Use minimum modulation.
(D) Reduce audio amplitude.

(126) Tests of survival craft radio equipment, EXCEPT EPIRBs and two-way radiotelephone equipment, must be conducted:

(A) At weekly intervals while the ship is at sea.
(B) Within 24 hours prior to departure when a test has not been conducted within a week of departure.
(C) Both A and B above.
(D) When required by the Commission.

(127) Each cargo ship of the United States which is equipped with a radiotelephone station for compliance with Part II of Title III of the Communications Act shall while being navigated outside of a harbor or port keep a continuous watch on:

(A) 2182 kHz.
(B) 156.8 Mhz.
(C) Both A and B.
(D) Cargo ships are exempt from radio watch regulations.

(128) When may you test a radiotelephone transmitter on the air?

(A) Between midnight and 6:00 AM local time.
(B) Only when authorized by the Commission.
(C) At any time as necessary to assure proper operation.
(D) After reducing transmitter power to 1 watt.

(129) What is the required daytime range of a radiotelephone station aboard a 900 ton ocean going cargo vessel?

(A) 25 miles.
(B) 50 miles.
(C) 150 miles.
(D) 500 miles.

(130) What do you do if the transmitter aboard your ship is operating off-frequency, overmodulating or distorting?

(A) Reduce to low power.
(B) Stop transmitting.
(C) Reduce audio volume level.
(D) Make a notation in station operating log.

(131) What is the authorized frequency for an on-board ship repeater for use

with a mobile transmitter operating at 467.750 MHz?

(A) 457.525 MHz.
(B) 467.775 MHz.
(C) 467.800 MHz.
(D) 467.825 MHz.

(132) Survival craft EPIRBs are tested:

(A) With a manually activated test switch.
(B) With a dummy load having the equivalent impedance of the antenna affixed to the EPIRB.
(C) With radiation reduced to a level not to exceed 25 microvolts per meter.
(D) All of the above.

(133) What safety signal call word is spoken three times, followed by the station call letters spoken three times, to announce a storm warning, danger to navigation, or special aid to navigation?

(A) PAN.
(B) MAYDAY.
(C) SECURITY.
(D) SAFETY.

(134) When should both the call sign and the name of the ship be mentioned during radiotelephone transmissions?

(A) At all times.
(B) During an emergency.
(C) When transmitting on 2182 kHz.
(D) Within 100 miles of any shore.

(135) How often is the auto alarm tested?

(A) During the 5-minute silent period.
(B) Monthly on 121.5 MHz using a dummy load.
(C) Weekly on frequencies other than the 2182 kHz distress frequency using a dummy antenna.
(D) Each day on 2182 kHz using a dummy antenna.

(136) One nautical mile is approximately equal to how many statute miles?

(A) 1.61 statute miles.
(B) 1.83 statute miles.
(C) 1.15 statute miles.

(D) 1.47 statute miles.

(137) A reserve power source must be able to power all radio equipment plus an emergency light system for how long?

(A) 24 hours.
(B) 12 hours.
(C) 8 hours.
(D) 6 hours.

(138) Frequencies used for portable communications on board ship:

(A) 9300-9500 MHz.
(B) 1636.5-1644 MHz.
(C) 2900-3100 MHz.
(D) 457.525-467.825 MHz.

(139) In the FCC rules the frequency band from 30 to 300 MHz is also known as:

(A) Very High Frequency (VHF).
(B) Ultra High Frequency (UHF).
(C) Medium Frequency (MF).
(D) High Frequency (HF).

(140) What channel must VHF-FM equipped vessels monitor at all times the station is operated?

(A) Channel 8; 156.4 MHz.
(B) Channel 16; 156.8 MHz.
(C) Channel 5A; 156.25 MHz.
(D) Channel 1A; 156.07 MHz.

(141) When testing is conducted within the 2170-2194 kHz and 156.75- 156.85 MHz. bands, transmissions should not continue for more than ___________ in any 15 minute period.

(A) 15 seconds.
(B) 1 minute.
(C) 5 minutes.
(D) No limitation.

(142) What emergency radio testing is required for cargo ships?

(A) Tests must be conducted weekly while ship is at sea.
(B) Full power carrier tests into dummy load.
(C) Specific gravity check in lead acid batteries, or voltage under

load for dry cell batteries.

(D) All of the above.

(143) The master or owner of a vessel must apply how many days in advance for an FCC ship inspection?

(A) 60 days.
(B) 30 days.
(C) 3 days.
(D) 24 hours.

(144) Marine transmitters should be modulated between:

(A) 75%-100%.
(B) 70%-105%
(C) 85%-100%
(D) 75%-120%

(145) What is a good practice when speaking into a microphone in a noisy location?

(A) Overmodulation.
(B) Change phase in audio circuits.
(C) Increase monitor audio gain.
(D) Shield microphone with hands.

(146) When pausing briefly for station copying message to acknowledge, say:

(A) BREAK.
(B) OVER.
(D) WILCO.
(D) STOP.

(147) Overmodulation is often caused by:

(A) Turning down audio gain control.
(B) Station frequency drift.
(C) Weather conditions.
(D) Shouting into microphone.

(148) To indicate a response is expected, say:

(A) WILCO.
(B) ROGER.
(C) OVER.
(D) BREAK.

(149) When all of a transmission has been received, say:

(A) ATTENTION.
(B) ROGER.
(C) RECEIVED.
(D) WILCO.

(150) What information must be included in a DISTRESS message?

(A) Name of vessel.
(B) Location.
(C) Type of distress and specifics of help requested.
(D) All of the above.

(151) The maritime MF radiotelephone silence periods begin at _______ and_______ minutes past the UTC hour.

(A) :15 , :45.
(B) :00 , :30.
(C) :20, :40.
(D) :05 , :35.

(152) A marine public coast station operator may not charge a fee for what type of communication?

(A) Port Authority transmissions.
(B) Storm updates.
(C) Distress.
(D) All of the above.

(153) Which of the following represent the first three letters of the phonetic alphabet?

(A) Alpha Bravo Charlie.
(B) Adam Baker Charlie.
(C) Alpha Baker Crystal.
(D) Adam Brown Chuck.

(154) Two way communications with both stations operating on the same frequency is:

(A) Radiotelephone.
(B) Duplex.
(C) Simplex.
(D) Multiplex.

(155) When a ship is sold:

(A) New owner must apply for a new license.
(B) FCC inspection of equipment is required.
(C) Old license is valid until it expires.
(D) Continue to operate; license automatically transfers with ownership.

(156) What is the second in order of priority?

(A) URGENT.
(B) DISTRESS.
(C) SAFETY.
(D) MAYDAY.

(157) Portable ship units, hand-helds or walkie-talkies used as an associated ship unit:

(A) Must operate with 1 watt and be able to transmit on Channel 16.
(B) May communicate only with the mother ship and other portable units and small boats belonging to mother ship.
(C) Must not transmit from shore or to other vessels.
(D) All of the above.

(158) The HF (high frequency) band is:

(A) 3 - 30 MHz
(B) 3 - 30 GHz
(C) 30 - 300 MHz
(D) 300 - 3000 MHz.

(159) Omega operates in what frequency band?

(A) Below 3 kHz.
(B) 3 - 30 kHz.
(C) 30 - 300 kHz.
(D) 300 -3000 kHz.

(160) Shipboard transmitters using F3E emission (FM voice) may not exceed what carrier power?

(A) 500 watts.
(B) 250 watts.
(C) 100 watts.
(D) 25 watts.

(161) Loran C operates in what frequency band?

(A) VHF; 30 -300 MHz.
(B) HF; 3 30 MHz.
(C) MF; 300 - 3000 kHz.
(D) LF; 30-300 kHz.

(162) What has most priority:

(A) URGENT.
(B) DISTRESS.
(C) SAFETY.
(D) SECURITY.

(163) When and how may Class A and B EPIRBs be tested?

(A) Within the first 5 minutes of the hour; tests not to exceed 3 audible sweeps or one second, whichever is longer.
(B) Within first 3 minutes of hour; tests not to exceed 30 seconds.
(C) Within first 1 minute of hour, test not to exceed 1 minute.
(D) At any time ship is at sea.

(164) When is the Silent Period on 2182 kHz, when only emergency communications may occur?

(A) One minute at the beginning of every hour and half hour.
(B) At all times.
(C) No designated period; silence is maintained only when a distress call is received.
(D) Three minutes at the beginning of every hour and half hour.

(165) What is the frequency range of UHF?

(A) 0.3 to 3 GHz.
(B) 0.3 to 3 MHz.
(C) 3 to 30 kHz.
(D) 30 to 300 MHz.

(166) A room temperature of + 30.0 degrees Celsius is equivalent to how many degrees Fahrenheit?

(A) 104.
(B) 83.
(C) 95.
(D) 86.

(167) Atmospheric noise or static is not a great problem:

(A) At frequencies below 20 MHz.
(B) At frequencies below 5 MHz.
(C) At frequencies above 1 MHz.
(D) At frequencies above 30 MHz.

(168) Frequencies which have substantially straight-line propagation characteristics similar to that of light waves are:

(A) Frequencies below 500 kHz.
(B) Frequencies between 500 kHz and 1,000 kHz.
(C) Frequencies between 1,000 kHz and 3,000 kHz.
(D) Frequencies above 50,000 kHz.

(169) In the International Phonetic Alphabet, the letters E, M, and S are represented by the words:

(A) Echo, Michigan, Sonar.
(B) Equator, Mike, Sonar.
(C) Echo, Mike, Sierra.
(D) Element, Mister, Scooter.

(170) What is the international radiotelephone distress call?

(A) "SOS, SOS, SOS; THIS IS;" followed by the call sign of the station (repeated 3 times).
(B) "MAYDAY, MAYDAY, MAYDAY; THIS IS;" followed by the call sign (or name, if no call sign assigned) of the mobile station in distress, spoken three times.
(C) For radiotelephone use, any words or message which will attract attention may be used.
(D) The alternating two tone signal produced by the radiotelephone alarm signal generator.

Element Three Questions

Element 3: General Radiotelephone. 76 questions concerning electronic fundamentals and techniques required to adjust, repair, and maintain radio transmitters and receivers at stations licensed by the FCC in the aviation, maritime, and international fixed public radio services. The minimum passing score is 57 questions answered correctly.

A General Radiotelephone Operator License (GROL) must also be held by the operator of certain aviation and maritime land radio stations; compulsory equipped ship radiotelephone stations transmitting with more than 1,500 watts of peak envelope power output; and voluntarily equipped ship stations transmitting with more than 1000 watts peak envelope power.

The questions asked on your Element 3 exam will be taken from the following 8 sections (3A through 3H), which contain a total of 728 questions. Your FCC exam will be constructed from the following algorithm:

Section 3A - Operating Procedures - 3 questions
Section 3B - Radio Wave Propagation - 3 questions
Section 3C - Radio Practice - 5 questions
Section 3D - Electrical Principles - 16 questions
Section 3E - Circuit Components - 13 questions
Section 3F - Practical Circuits - 22 questions
Section 3G - Signals and Emissions - 9 questions
Section 3H - Antennas and Feed Lines - 5 questions

Test 3-A Element Three Operating Procedures

3A1
What is facsimile?

A. The transmission of characters by radioteletype that form a picture when printed
B. The transmission of still pictures by slow-scan television
C. The transmission of video by television
D. The transmission of printed pictures for permanent display on paper

3A2
What is the modern standard scan rate for a facsimile picture transmitted by a radio station?

A. The modern standard is 240 lines per minute
B. The modern standard is 50 lines per minute
C. The modern standard is 150 lines per second
D. The modern standard is 60 lines per second

3A3
What is the approximate transmission time for a facsimile picture transmitted by a radio station?

A. Approximately 6 minutes per frame at 240 lines per minute
B. Approximately 3.3 minutes per frame at 240 lines per minute
C. Approximately 6 seconds per frame at 240 lines per minute
D. 1/60 second per frame at 240 lines per minute

3A4
What is the term for the transmission of printed pictures by radio for the purpose of a permanent display?

A. Television
B. Facsimile
C. Xerography
D. ACSSB

3A5
In facsimile, how are variations in picture brightness and darkness converted into voltage variations?

A. With an LED
B. With a Hall-effect transistor
C. With a photodetector
D. With an optoisolator

3A6
What is an ascending pass for a satellite?

A. A pass from west to east
B. A pass from east to west
C. A pass from south to north
D. A pass from north to south

3A7
What is a descending pass for a satellite?

A. A pass from north to south
B. A pass from west to east
C. A pass from east to west
D. A pass from south to north

3A8
What is the period of a satellite?

A. An orbital arc that extends from 60 degrees west longitude to 145 degrees west longitude
B. The point on an orbit where satellite height is minimum
C. The amount of time it takes for a satellite to complete one orbit
D. The time it takes a satellite to travel from perigee to apogee

3A9
What is a linear transponder?

A. A repeater that passes only linear or binary signals
B. A device that receives and retransmits signals of any mode in a certain passband
C. An amplifier for SSB transmissions
D. A device used to change a FM emission to an AM emission

3A10
What are the two basic types of linear transponders used in satellites?

A. Inverting and non-inverting
B. Geostationary and elliptical
C. Phase 2 and Phase 3
D. Amplitude modulated and frequency modulated

3A11
Why does the downlink frequency appear to vary by several kHz during a low-earth-orbit satellite pass?

A. The distance between the satellite and ground station is changing, causing the Kepler effect
B. The distance between the satellite and ground station is changing, causing the Bernoulli effect
C. The distance between the satellite and ground station is changing, causing the Boyles' law effect
D. The distance between the satellite and ground station is changing, causing the Doppler effect

3A12
Why does the received signal from a satellite stabilized by a computer-pulsed electromagnet exhibit a fairly rapid pulsed fading effect?

A. Because the satellite is rotating
B. Because of ionospheric absorption
C. Because of the satellite's low orbital altitude
D. Because of the Doppler effect

3A13
What type of antenna can be used to minimize the effects of spin modulation and Faraday rotation?

A. A nonpolarized antenna
B. A circularly polarized antenna
C. An isotropic antenna
D. A log-periodic dipole array

3A14
What is blanking in a video signal?

A. Synchronization of the horizontal and vertical sync-pulses
B. Turning off the scanning beam while it is traveling from right to left and from bottom to top
C. Turning off the scanning beam at the conclusion of a transmission
D. Transmitting a black and white test pattern

3A15
What is the standard video voltage level between the sync tip and the whitest white at TV camera outputs and modulator inputs?

A. 1 volt peak-to-peak
B. 120 IEEE units
C. 12 volts DC
D. 5 volts RMS

3A16
What is the standard video level, in percent PEV, for black?

A. 0%
B. 12.5%
C. 70%
D. 100%

3A17
What is the standard video level, in percent PEV, for white?

A. 0%
B. 12.5%
C. 70%
D. 100%

3A18
What is the standard video level, in percent PEV, for blanking?

A. 0%
B. 12.5%
C. 75%
D. 100%

3A19
A 25 MHz. amplitude modulated transmitter's actual carrier frequency is 25.00025 MHz without modulation and is 24.99950 MHz when modulated. What statement is true?

A. If the allowed frequency tolerance is 0.001%, this is an illegal transmission
B. If the allowed frequency tolerance is 0.002%, this is an illegal transmission
C. Modulation should not change carrier frequency
D. If the authorized frequency tolerance is 0.005% for the 25 MHz band this transmitter is operating legally

3A20
If you are listening to an FM radio station at 100.6 MHz. on a car radio when an airplane in the vicinity is transmitting at 121.2 MHz. and your car radio receives interference the possible problem could be:

A. Improper shielding in receive
B. Poor "Q" receiver
C. Image frequency
D. Intermodulation or coupling

3A21
One nautical mile is equal to how many statute miles?

A. 1.5
B. 8.3
C. 1.73
D. 1.15

3A22
1.73 nautical miles equals how many statute miles?

A. 2
B. 1.5
C. 1.73
D. 1

3A23
Solder is:

A. 50% lead 50% tin
B. 40% lead 60% tin
C. 60% lead 40% tin
D. 70% lead 30% tin

3A24
What ferrite device can be used instead of a duplexer to isolate a microwave transmitter and receiver when both are connected to the same antenna?

A. Isolator
B. Circulator
C. Magnetron
D. Simplex

3A25
The ILS localizer measures what deviation of an aircraft?

A. Horizontal
B. Vertical
C. Ground speed
D. Distance between aircraft

3A26
3:00 PM Central Standard Time is:

A. 1000 UTC
B. 2100 UTC
C. 1800 UTC
D. 0300 UTC

3A27
10 statute miles per hour equals how many knots?

A. 11.5
B. 8.7
C. 5
D. 3

3A28
Which of the following is an acceptable method of solder removal from holes in a printed board?

A. Compressed air
B. Toothpick
C. Soldering iron and a suction device
D. Power drill

3A29
100 statute miles equals how many nautical miles?

A. 87
B. 108
C. 173
D. 13

3A30
6:00 PM PST is equal to what time in UTC?

A. 0200
B. 1800
C. 2300
D. 1300

3A31
An Auto Alarm signal consists of two sine wave audio tones transmitted alternately at what frequencies?

A. 121.5 and 243 MHz.
B. 500 and 1000 kHz.
C. 1300 and 220 kHz.
D. none of the above

3A32
What is NOT a good soldering practice in electronic circuits?

A. Use adequate heat
B. Clean parts sufficiently
C. Prevent corrosion by never using flux
D. Be certain parts do not move while solder is cooling

3A33
Waveguides are not utilized at VHF or UHF frequencies because:

A. Characteristic impedance
B. Resistance to high frequency waves
C. Large dimensions of waveguide are not practical
D. Grounding problems

3A34
2300 UTC time is:

A. 2 PM CST
B. 3 PM PST
C. 10 AM EST
D. 6 AM EST

3A35
One statute mile equals how many nautical miles?

A. 3.8
B. 1.5
C. 0.87
D. 0.7

3A36
When soldering electronic circuits be sure to:

A. Use sufficient heat
B. Use maximum heat
C. Heat wires until sweating begins
D. Use minimum solder

3A37
2.3 statute miles equals how many nautical miles?

A. 2
B. 1.5
C. 1.73
D. 1

3A38
What is the purpose of flux?

A. Removes oxides from surfaces to be joined
B. Prevents oxidation during soldering
C. Acid cleans printed circuit connections
D. Both a and b

3A39
Which of these will be useful for insulation at UHF?

A. Rubber
B. Mica
C. Wax inpregnated paper
D. Lead

3A40
The condition of a storage battery is determined with a (n):

A. Hygrometer
B. Manometer
C. FET
D. Hydrometer

Test 3-B Element Three
Radio Wave Propagation

3B1
What is a selective fading effect?

A. A fading effect caused by small changes in beam heading at the receiving station
B. A fading effect caused by phase differences between radio wave components of the same transmission, as experienced at the receiving station
C. A fading effect caused by large changes in the height of the ionosphere, as experienced at the receiving station
D. A fading effect caused by time differences between the receiving and transmitting stations

3B2
What is the propagation effect called when phase differences between radio wave components of the same transmission are experienced at the recovery station?

A. Faraday rotation
B. Diversity reception
C. Selective fading
D. Phase shift

3B3
What is the major cause of selective fading?

A. Small changes in beam heading at the receiving station
B. Large changes in the height of the ionosphere, as experienced at the receiving station
C. Time differences between the receiving and transmitting stations
D. Phase differences between radio wave components of the same transmission, as experienced at the receiving station

3B4
Which emission modes suffer the most from selective fading?

A. CW and SSB
B. FM and double sideband AM
C. SSB and image
D. SSTV and CW

3B5
How does the bandwidth of the transmitted signal affect selective fading?

A. It is more pronounced at wide bandwidths
B. It is more pronounced at narrow bandwidths
C. It is equally pronounced at both narrow and wide bandwidths
D. The receiver bandwidth determines the selective fading effect

3B6
Why does the radio-path horizon distance exceed the geometric horizon?

A. E-layer skip
B. D-layer skip
C. Auroral skip
D. Radio waves may be bent

3B7
How much farther does the radio-path horizon distance exceed the geometric horizon?

A. By approximately 15% of the distance
B. By approximately twice the distance
C. By approximately one-half the distance
D. By approximately four times the distance

3B8
To what distance is VHF propagation ordinarily limited?

A. Approximately 1000 miles
B. Approximately 500 miles
C. Approximately 1500 miles
D. Approximately 2000 miles

3B9
What propagation condition is usually indicated when a VHF signal is received from a station over 500 miles away?

A. D-layer absorption
B. Faraday rotation
C. Tropospheric ducting
D. Moonbounce

3B10
What happens to a radio wave as it travels in space and collides with other particles?

A. Kinetic energy is given up by the radio wave
B. Kinetic energy is gained by the radio wave
C. Aurora is created
D. Nothing happens since radio waves have no physical substance

3B11
When the earth's atmosphere is struck by a meteor, a cylindrical region of free electrons is formed at what layer of the ionosphere?

A. The F1 layer
B. The E layer
C. The F2 layer
D. The D layer

3B12
What is transequatorial propagation?

A. Propagation between two points at approximately the same distance north and south of the magnetic equator
B. Propagation between two points on the magnetic equator
C. Propagation between two continents by way of ducts along the magnetic equator
D. Propagation between any two stations at the same latitude

3B13
What is the maximum range for signals using transequatorial propagation?

A. About 1,000 miles
B. About 2,500 miles
C. About 5,000 miles
D. About 7,500 miles

3B14
What is the best time of day for transequatorial propagation?

A. Morning
B. Noon
C. Afternoon or early evening
D. Transequatorial propagation only works at night

3B15
What is knife-edge diffraction?

A. Allows normally line-of-sight signals to bend around sharp edges, mountain ridges, buildings and other obstructions
B. Arcing in sharp bends of conductors
C. Phase angle image rejection
D. Line-of-sight signals causing distortion to other signals

3B16
The average range for VHF communications is:

A. 5 miles
B. 15 miles
C. 30 miles
D. 100 miles

3B17
The bending of radio waves passing over the top or a mountain range that disperses a weak portion of the signal behind the mountain is:

A. Eddy-current phase effect
B. Knife-edge diffraction
C. Shadowing
D. Mirror refraction effect

3B18
Knife-edge refraction:

A. Is the bending of UHF frequency radio waves around a building, mountain or obstruction
B. Causes the velocity of wave propagation to be different than original wave
C. Both a and b above
D. Attenuates UHF signals

3B19
If the elapsed time for a radar echo is 62 microseconds what is the distance in nautical miles to the object?

A. 5
B. 87
C. 37
D. 11.5

3B20
What is the wavelength of a signal at 500 MHz?

A. 0.062 cm
B. 6 meters
C. 60 cm
D. 60 meters

3B21
The radar range in nautical miles to an object can be found by measuring the elapsed time during a radar pulse and dividing this quantity by:

A. 0.87 seconds
B. 1.15 microseconds
C. 12.36 microseconds
D. 1.73 microseconds

3B22
The band of frequencies least susceptible to atmospheric ncise and interference is:

A. 30 - 300 kHz.
B. 300 - 3000 kHz.
C. 3 - 30 MHz.
D. 300 - 3000 MHz.

Test 3-C Element Three
Radio Practice

3C1
What is a frequency standard?

A. A well-known (standard) frequency used for transmitting certain messages
B. A device used to produce a highly accurate reference frequency
C. A device for accurately measuring frequency to within 1 Hz
D. A device used to generate wide-band random frequencies

3C2
What is a frequency-marker generator?

A. A device used to produce a highly accurate reference frequency
B. A sweep generator
C. A broadband white noise generator
D. A device used to generate wide-band random frequencies

3C3
How is a frequency-marker generator used?

A. In conjunction with a grid-dip meter
B. To provide reference points on a receiver dial
C. As the basic frequency element of a transmitter
D. To directly measure wavelength

3C4
What is a frequency counter?

A. A frequency measuring device
B. A frequency marker generator
C. A device that determines whether or not a given frequency is in use before automatic transmissions are made
D. A broadband white noise generator

3C5
How is a frequency counter used?

A. To provide reference points on an analog receiver dial
B. To generate a frequency standard
C. To measure the deviation in an FM transmitter
D. To measure frequency

3C6
What is the most the actual transmitter frequency could differ from a reading of 156,520,000-Hertz on a frequency counter with a time base accuracy of +/- 1.0 ppm?

A. 165.2 Hz
B. 15.652 kHz
C. 156.52 Hz
D. 1.4652 MHz

3C7
What is the most the actual transmitter frequency could differ from a reading of 156,520,000-Hertz on a frequency counter with a time base accuracy of +/- 0.1 ppm?

A. 15.652 Hz
B. 0.1 MHz
C. 1.4652 Hz
D. 1.5652 kHz

3C8
What is the most the actual transmitter frequency could differ from a reading of 156,520,000-Hertz on a frequency counter with a time base accuracy of +/- 10 ppm?

A. 146.52 Hz
B. 10 Hz
C. 156.52 kHz
D. 1565.20 Hz

3C9
What is the most the actual transmitter frequency could differ from a reading of 462,100,000-Hertz on a frequency counter with a time base accuracy of +/- 1.0 ppm?

A. 46.21 MHz
B. 10 Hz
C. 1.0 MHz
D. 462.1 Hz

3C10
What is the most the actual transmit frequency could differ from a reading of 462,100,000-Hertz on a frequency counter with a time base accuracy of +/- 0.1ppm?

A. 46.21 Hz
B. 0.1 MHz
C. 462.1 Hz
D. 0.2 MHz

3C11
What is the most the actual transmit frequency could differ from a reading of 462,100,000-Hertz on a frequency counter with a time base accuracy of +/- 10 ppm?

A. 10 MHz
B. 10 Hz
C. 4621 Hz
D. 462.1 Hz

3C12
What is a dip-meter?

A. A field strength meter
B. An SWR meter
C. A variable LC oscillator with metered feedback current
D. A marker generator

3C13
Why is a dip-meter used by many technicians?

A. It can measure signal strength accurately
B. It can measure frequency accurately
C. It can measure transmitter output power accurately
D. It can give an indication of the resonant frequency of a circuit

3C14
How does a dip-meter function?

A. Reflected waves at a specific frequency desensitize the detector coil
B. Power coupled from an oscillator causes a decrease in metered current
C. Power from a transmitter cancels feedback current
D. Harmonics of the oscillator cause an increase in resonant circuit Q

3C15
What two ways could a dip-meter be used in a radio station?

A. To measure resonant frequency of antenna traps and to measure percentage of modulation
B. To measure antenna resonance and to measure percentage of modulation
C. To measure antenna resonance and to measure antenna impedance
D. To measure resonant frequency of antenna traps and to measure a tuned circuit resonant frequency

3C16
What types of coupling occur between a dip-meter and a tuned circuit being checked?

A. Resistive and inductive
B. Inductive and capacitive
C. Resistive and capacitive
D. Strong field

3C17
How tight should the dip-meter be coupled with the tuned circuit being checked?

A. As loosely as possible, for best accuracy
B. As tightly as possible, for best accuracy
C. First loose, then tight, for best accuracy
D. With a soldered jumper wire between the meter and the circuit to be checked, for best accuracy

3C18
What happens in a dip-meter when it is too tightly coupled with the tuned circuit being checked?

A. Harmonics are generated
B. A less accurate reading results
C. Cross modulation occurs
D. Intermodulation distortion occurs

3C19
What factors limit the accuracy, frequency response, and stability of an oscilloscope?

A. Sweep oscillator quality and deflection amplifier bandwidth
B. Tube face voltage increments and deflection amplifier voltage
C. Sweep oscillator quality and tube face voltage increments
D. Deflection amplifier output impedance and tube face frequency increments

3C20
What factors limit the accuracy, frequency response, and stability of a D'Arsonval movement type meter?

A. Calibration, coil impedance and meter size
B. Calibration, series resistance and electromagnet current
C. Coil impedance, electromagnet voltage and movement mass
D. Calibration, mechanical tolerance and coil impedance

3C21
What factors limit the accuracy, frequency response, and stability of a frequency counter?

A. Number of digits in the readout, speed of the logic and time base stability
B. Time base accuracy, speed of the logic and time base stability
C. Time base accuracy, temperature coefficient of the logic and time base stability
D. Number of digits in the readout, external frequency reference and temperature coefficient of the logic

3C22
How can the frequency response of an oscilloscope be improved?

A. By using a triggered sweep and a crystal oscillator as the time base
B. By using a crystal oscillator as the time base and increasing the vertical sweep rate
C. By increasing the vertical sweep rate and the horizontal amplifier frequency response
D. By increasing the horizontal sweep rate and the vertical amplifier frequency response

3C23
How can the accuracy of a frequency counter be improved?

A. By using slower digital logic
B. By improving the accuracy of the frequency response
C. By increasing the accuracy of the time base
D. By using faster digital logic

3C24
What is the name of the condition that occurs when the signals of two transmitters in close proximity mix together in one or both of their final amplifiers, and unwanted signals at the sum and difference frequencies of the original transmissions are generated?

A. Amplifier desensitization
B. Neutralization
C. Adjacent channel interference
D. Intermodulation interference

3C25
How does intermodulation interference between two transmitters usually occur?

A. When the signals from the transmitters are reflected out of phase from airplanes passing overhead
B. When they are in close proximity and the signals mix in one or both of their final amplifiers
C. When they are in close proximity and the signals cause feedback in one or both of their final amplifiers
D. When the signals from the transmitters are reflected in phase from airplanes passing overhead

3C26
How can intermodulation interference between two transmitters in close proximity often be reduced or eliminated?

A. By using a Class C final amplifier with high driving power
B. By installing a terminated circulator or ferrite isolator in the feed line to the transmitter and duplexer
C. By installing a band-pass filter in the antenna feed line
D. By installing a low-pass filter in the antenna feed line

3C27
What can occur when a non-linear amplifier is used with a single-sideband phone transmitter?

A. Reduced amplifier efficiency
B. Increased intelligibility
C. Sideband inversion
D. Distortion

3C28
How can even-order harmonics be reduced or prevented in transmitter amplifier design?

A. By using a push-push amplifier
B. By using a push-pull amplifier
C. By operating class C
D. By operating class AB

3C29
What is receiver desensitizing?

A. A burst of noise when the squelch is set too low
B. A burst of noise when the squelch is set too high
C. A reduction in receiver sensitivity because of a strong signal on a nearby frequency
D. A reduction in receiver sensitivity when the AF gain control is turned down

3C30
What is the term used to refer to the reduction of receiver gain caused by the signals of a nearby station transmitting in the same frequency band?

A. Desensitizing
B. Quieting
C. Cross modulation interference
D. Squelch gain rollback

3C31
What is the term used to refer to a reduction in receiver sensitivity caused by unwanted high-level adjacent channel signals?

A. Intermodulation distortion
B. Quieting
C. Desensitizing
D. Overloading

3C32
How can receiver desensitizing be reduced?

A. Ensure good RF shielding between the transmitter and receiver
B. Increase the transmitter audio gain
C. Decrease the receiver squelch gain
D. Increase the receiver bandwidth

3C33
What is cross-modulation interference?

A. Interference between two transmitters of different modulation type
B. Interference caused by audio rectification in the receiver preamp
C. Harmonic distortion of the transmitted signal
D. Modulation from an unwanted signal is heard in addition to the desired signal

3C34
What is the term used to refer to the condition where the signals from a very strong station are superimposed on other signals being received?

A. Intermodulation distortion
B. Cross-modulation interference
C. Receiver quieting
D. Capture effect

3C35
How can cross-modulation in a receiver be reduced?

A. By installing a filter at the receiver
B. By using a better antenna
C. By increasing the receiver's RF gain while decreasing the AF gain
D. By adjusting the pass-band tuning

3C36
What is the result of cross-modulation?

A. A decrease in modulation level of transmitted signals
B. Receiver quieting
C. The modulation of an unwanted signal is heard on the desired signal
D. Inverted sidebands in the final stage of the amplifier

3C37
What is the capture effect?

A. All signals on a frequency are demodulated by an FM receiver
B. All signals on a frequency are demodulated by an AM receiver
C. The loudest signal received is the only demodulated signal
D. The weakest signal received is the only demodulated signal

3C38
What is the term used to refer to the reception blockage of one FM-phone signal by another FM-phone signal?

A. Desensitization
B. Cross-modulation interference
C. Capture effect
D. Frequency discrimination

3C39
With which emission type is the capture-effect most pronounced?

A. FM
B. SSB
C. AM
D. CW

3C40
How does a spectrum analyzer differ from a conventional time-domain oscilloscope?

A. The oscilloscope is used to display electrical signals while the spectrum analyzer is used to measure ionospheric reflection
B. The oscilloscope is used to display electrical signals in the frequency domain while the spectrum analyzer is used to display electrical signals in the time domain
C. The oscilloscope is used to display electrical signals in the time domain while the spectrum analyzer is used to display electrical signals in the frequency domain
D. The oscilloscope is used for displaying audio frequencies and the spectrum analyzer is used for displaying radio frequencies

3C41
What does the horizontal axis of a spectrum analyzer display?

A. Amplitude
B. Voltage
C. Resonance
D. Frequency

3C42
What does the vertical axis of a spectrum analyzer display?

A. Amplitude
B. Duration
C. Frequency
D. Time

3C43
What test instrument can be used to display spurious signals in the output of a radio transmitter?

A. A spectrum analyzer
B. A wattmeter
C. A logic analyzer
D. A time-domain reflectometer

3C44
What test instrument is used to display intermodulation distortion products from an SSB transmitter?

A. A wattmeter
B. A spectrum analyzer
C. A logic analyzer
D. A time-domain reflectometer

3C45
What advantage does a logic probe have over a voltmeter for monitoring logic states in a circuit?

A. A logic probe has fewer leads to connect to a circuit than a voltmeter
B. A logic probe can be used to test analog and digital circuits
C. A logic probe can be powered by commercial AC lines.
D. A logic probe is smaller and shows a simplified readout

3C46
What piece of test equipment can be used to directly indicate high and low logic states?

A. A galvanometer
B. An electroscope
C. A logic probe
D. A Wheatstone bridge

3C47
What is a logic probe used to indicate?

A. A short-circuit fault in a digital-logic circuit
B. An open-circuit failure in a digital-logic circuit
C. A high-impedance ground loop
D. High and low logic states in a digital-logic circuit

3C48
What piece of test equipment besides an oscilloscope can be used to indicate pulse conditions in a digital-logic circuit?

A. A logic probe
B. A galvanometer
C. An electroscope
D. A Wheatstone bridge

3C49
What is one of the most significant problems you might encounter when you try to receive signals with a mobile station?

A. Ignition noise
B. Doppler shift
C. Radar interference
D. Mechanical vibrations

3C50
What is the proper procedure for suppressing electrical noise in a mobile station?

A. Apply shielding and filtering where necessary
B. Insulate all plane sheet metal surfaces from each other
C. Apply antistatic spray liberally to all non-metallic surfaces
D. Install filter capacitors in series with all DC wiring

3C51
How can ferrite beads be used to suppress ignition noise?

A. Install them in the resistive high voltage cable every 2 years
B. Install them between the starter solenoid and the starter motor
C. Install them in the primary and secondary ignition leads
D. Install them in the antenna lead to the radio

3C52
How can conducted and radiated noise caused by an alternator be suppressed?

A. By installing filter capacitors in series with the DC power lead and by installing a blocking capacitor in the field lead
B. By connecting the radio's power leads to the battery by the longest possiblepath and by installing a blocking capacitor in series with the positive lead
C. By installing a high pass filter in series with the radio's power lead to the vehicle's electrical system and by installing a low-pass filter in parallel with the field lead
D. By connecting the radio's power leads directly to the battery and by installing coaxial capacitors in the alternator leads

3C53
What is a major cause of atmospheric static?

A. Sunspots
B. Thunderstorms
C. Airplanes
D. Meteor showers

3C54
How can you determine if a line-noise interference problem is being generated within a building?

A. Check the power-line voltage with a time-domain reflectometer
B. Observe the AC wave form on an oscilloscope
C. Turn off the main circuit breaker and listen on a battery-operated radio
D. Observe the power-line voltage on a spectrum analyzer

3C55
What causes receiver desensitizing?

A. Audio gain adjusted too low
B. Squelch gain adjusted too high
C. The presence of a strong signal on a nearby frequency
D. Squelch gain adjusted too low

3C56
How can alternator whine be minimized?

A. By connecting the radio's power leads to the battery by the longest possible path
B. By connecting the radio's power leads to the battery by the shortest possible path
C. By installing a high pass filter in series with the radio's DC power lead to the vehicle's electrical system
D. By installing filter capacitors in series with the DC power lead

3C57
An electrical relay:

A. Is a current limiting device
B. Is a device used for supplying 3 or more voltages to a circuit
C. Is concerned mainly with HF audio amplifiers
D. Provides either an open or closed circuit to some position of a device

3C58
A high standing wave ratio on a transmission line can be caused by:

A. Excessive modulation
B. An increase in output power
C. Detuned antenna coupling
D. Poor B+ regulation

3C59
The best insulation at UHF is:

A. Black rubber
B. Bakelite
C. Paper
D. Mica

3C60
The electrolyte in a lead storage battery is:

A. Potassium hydrate
B. Pure spongy lead
C. Sulphuric acid
D. Iron oxide

3C61
A dynamotor is used to:

A. Step up AC voltage
B. Step down AC voltage
C. Step down DC voltage
D. Step up DC voltage

3C62
The SHF band is:

A. 3000 to 30,000 MHz
B. Above 300,000 MHz
C. 300 to 3000 MHz
D. 30,000 to 300,000 MHz

3C63
When soldering electrical connections, care should be taken not to:

A. Heat both connections
B. Move the wire until solder has solidified
C. Have solder on the connections
D. Use rosin core solder

3C64
If there are too many harmonics from a transmitter, check the:

A. Coupling
B. Tuning of circuits
C. Shielding
D. Any of the above

3C65
One piece of equipment to indicate neutralization is:

A. A neon bulb
B. A tachometer
C. Wave trap
D. Filter

3C66
Interference to radio receivers in automobiles can be reduced by:

A. Connecting resistances in series with the spark plugs
B. Using heavy conductors between the starting battery and the starting motor
C. Connecting resistances in series with the starting battery
D. Grounding the negative side of the starting battery

3C67
Magnetron oscillators are used for:

A. Generating SHF signals
B. Multiplexing
C. Generating rich harmonics
D. FM demodulation

3C68
Normally in a mobile radio installation the E output of a dynamotor is:

A. Adjusted by armature rheostat
B. Adjusted by field rheostat
C. Not adjustable
D. Adjusted by battery rheostat

3C69
If a shunt motor, running with a load, has its shunt field opened, how would this affect the speed of the motor?

A. Slow down
B. Stop suddenly
C. Speed up then slow down
D. Unaffected

3C70
The state of charge of a Lead Acid storage cell is determined by:

A. Hydrometer reading
B. Its open circuit voltage
C. Its short circuit discharge current
D. Ohmeter

3C71
An absortion wave meter is useful in measuring:

A. Field strength
B. Output frequencies to conform with FCC tolerance
C. Standing wave frequencies
D. The resonant frequency of LC tank circuit

3C72
If a power supply fails, a new high vacuum tube glows red-hot, smoke curls from around the filter choke:

A. The transformer secondary has shorted
B. The filter choke has an open turn
C. One of the filter capacitors has shorted
D. A bleeder resitor has opened

3C73
Velocity of radio wave propogation in free space does what?

A. Varies in speed
B. Is constant
C. Is same as speed of light
D. Both b and c above

3C74
In storing a fully charged battery (lead-acid) for a long period of time (8 to 12 months) one should:

A. Drain and flush out the electrolyte then refill with distilled water
B. Keep it in a warm place
C. Jar it at least once a week to prevent sulfation
D. Maintain a constant trickle charge

3C75
A screen grid resistor burns out after smoking. First thing to check would be:

A. Short in screen grid by-pass capacitor
B. An open turn in plate coil
C. Short in B+ by-pass capacitor
D. Open grid leak resistor

3C76
Neglecting line losses, the RMS voltage along an RF transmission line having no standing waves:

A. Is equal to the impedance
B. Is one-half of the surge impedance
C. Is the product of the surge impedance times the line current
D. Varies sinusodially along the line

3C77
Waveguides are:

A. A hollow tube that carries RF
B. Solid conductor of RF
C. Coaxial cables
D. Copper wire

3C78
When a vacuum tube operates at VHF or higher as compared to lower frequencies:

A. Transit time of electrons becomes important
B. It is necessary to make larger components
C. It is necessary to increase grid spacing
D. Only pentode is satisfactory

3C79
The output voltage of a separately excited AC generator (running at a constant speed) could be controlled by:

A. Changing the dielectric constant of the armature
B. Changing the material of the armature
C. The field current
D. Operating the generator in a no-load condition

3C80
In radio circuits, the component most apt to break down is the:

A. Resistor
B. Crystal
C. Transformer
D. Wiring

3C81
A circulator:

A. Cools DC motors during heavy loads
B. Allows two or more antennas to feed one transmitter
C. Allows one antenna to feed two separate microwave transmitters and receivers at the same time
D. Insulates UHF frequencies on transmission lines

3C82
Static and interference from motors can be eliminated by:

A. Grounding the battery with a 2" copper strip
B. Installing faraday shields around connectors
C. Installing RF chokes across power line to ground
D. Installing bypass capactitators from the power line to grounded parts

3C83
Transmission line shielding is grounded:

A. At the input only
B. At both input and output
C. At the output only
D. If the antenna is a marconi design

3C84
Motorboating (low frequency oscillations) in an amplfier can be stopped by:

A. Grounding the screen grid
B. Bypassing the screen grid resistor with a .1 mfd capacitor
C. Connect a capacitor between the B+ lead and ground
D. Grounding the plate

3C85
Auto interference to radio reception can be eliminated by:

A. Installing resistive spark plugs
B. Installing capacitive spark plugs
C. Installing resistors in series with the spark plugs
D. installing two copper-braid ground strips

3C86
A DC series motor speed is affected by:

A. The load
B. The ripple frequency
C. The number of brushes
D. Stray RF fields

3C87
An antenna radiates a primary signal of 500 watts output. If there is a 2nd harmonic output of 0.5 watt, what attenuation of the 2nd harmonic has occured?

A. 3 dB
B. 10 dB
C. 20 dB
D. 30 dB

3C88
On runway approach, an ILS Localizer shows:

A. Deviation left or right of runway center line
B. Deviation up or down from ground speed
C. Deviation percentage from authorized ground speed
D. wind speed along runway

3C89
Where should a heat sink or heat shunt (such as a clamp or pliers) be placed when desoldering an IC chip?

A. On the tip of the soldering iron
B. Between the IC chip terminal and the soldering iron
C. Place clamp across all terminals
D. Between solder iron tip and the solder connection

3C90
What type of antenna system allows you to receive and transmit at the same time in both directions?

A. Simplex
B. Duplex
C. Multiplex
D. Digital diplex

3C91
A microwave device which allows RF energy to pass through in one direction with very little loss but absorbs RF power in the opposite direction:

A. Circulator
B. Wave trap
C. Multiplex
D. Isolator

3C92
The output voltage of a separately excited AC generator with constant frequency is dependant upon the adjustment of:

A. Iron or ferrite choke core
B. Field current
C. Brush position
D. Phase angle

3C93
What radio navigation aid determines the distance from a transponder beacon by measuring the length of time the radio signal took to travel to the receiver?

A. Radar
B. Loran C
C. Distance Marking (DM)
D. Distance Measuring Equipment (DME)

3C94
Which of the following is a feature of an instrument landing system (ILS)?

A. Localizer: shows aircraft deviation horizontally from center of runway
B. Glide slope (or glide path): shows aircraft vertical altitude of an aircraft during landing
C. Provides communication to aircraft
D. Both a and b

3C95
What is azimuth on a radar antenna?

A. Diameter
B. Degrees elevation
C. Degrees horizon
D. Impedance

3C96
Why do we tin component leads?

A. It helps to oxidize the wires
B. Prevents resting of circuit board
C. Decreases heating time and aids in connection
D. Provides ample wetting for good connection

3C97
What is the purpose of using a small amount of solder on the tip of a soldering iron just prior to making a connection?

A. Removes oxidation
B. Burns up flux
C. Increases solder temperature
D. Aids in wetting the wires

Test 3-D Element Three Electrical Principles

3D1
What is reactive power?

A. Wattless, non-productive power
B. Power consumed in wire resistance in an inductor
C. Power lost because of capacitor leakage
D. Power consumed in circuit Q

3D2
What is the term for an out-of-phase, non-productive power associated with inductors and capacitors?

A. Effective power
B. True power
C. Peak envelope power
D. Reactive power

3D3
What is the term for energy that is stored in an electromagnetic or electrostatic field?

A. Potential energy
B. Amperes-joules
C. Joules-coulombs
D. Kinetic energy

3D4
What is responsible for the phenomenon when voltages across reactances in series can often be larger than the voltages applied to them?

A. Capacitance
B. Resonance
C. Conductance
D. Resistance

3D5
What is resonance in an electrical circuit?

A. The highest frequency that will pass current
B. The lowest frequency that will pass current
C. The frequency at which capacitive reactance equals inductive reactance
D. The frequency at which power factor is at a minimum

3D6
Under what conditions does resonance occur in an electrical circuit?

A. When the power factor is at a minimum
B. When inductive and capacitive reactances are equal
C. When the square root of the sum of the capacitive and inductive reactances is equal to the resonant frequency
D. When the square root of the product of the capacitive and inductive reactances is equal to the resonant frequency

3D7
What is the term for the phenomena which occurs in an electrical circuit when the inductive reactance equals the capacitive reactance?

A. Reactive quiescence
B. High Q
C. Reactive equilibrium
D. Resonance

3D8
What is the approximate magnitude of the impedance of a series R-L-C circuit at resonance?

A. High, as compared to the circuit resistance
B. Approximately equal to the circuit resistance
C. Approximately equal to XL
D. Approximately equal to XC

3D9
What is the approximate magnitude of the impedance of a parallel R-L-C circuit at resonance?

A. Approximately equal to the circuit resistance
B. Approximately equal to XL
C. Low, as compared to the circuit resistance
D. Approximately equal to XC

3D10
What is the characteristic of the current flow in a series R-L-C circuit at resonance?

A. It is at a minimum
B. It is at a maximum
C. It is DC
D. It is zero

3D11
What is the characteristic of the current flow in a parallel R-L-C circuit at resonance?

A. The current circulating in the parallel elements is at a minimum
B. The current circulating in the parallel elements is at a maximum
C. The current circulating in the parallel elements is DC
D. The current circulating in the parallel elements is zero

3D12
What is the skin effect?

A. The phenomenon where RF current flows in a thinner layer of the conductor, close to the surface, as frequency increases
B. The phenomenon where RF current flows in a thinner layer of the conductor, close to the surface, as frequency decreases
C. The phenomenon where thermal effects on the surface of the conductor increasethe impedance
D. The phenomenon where thermal effects on the surface of the conductor decrease the impedance

3D13
What is the term for the phenomenon where most of an RF current flows along the surface of the conductor?

A. Layer effect
B. Seeburg Effect
C. Skin effect
D. Resonance

3D14
Where does practically all of the RF current flow in a conductor?

A. Along the surface
B. In the center of the conductor
C. In the magnetic field around the conductor
D. In the electromagnetic field in the conductor center

3D15
Why does practically all of an RF current flow within a few thousandths-of-an-inch of the conductor's surface?

A. Because of skin effect
B. Because the RF resistance of the conductor is much less than the DC resistance
C. Because of heating of the metal at the conductor's interior
D. Because of the AC-resistance of the conductor's self inductance

3D16
Why is the resistance of a conductor different for RF current than for DC?

A. Because the insulation conducts current at radio frequencies
B. Because of the Heisenbürg Effect
C. Because of skin effect
D. Because conductors are non-linear devices

3D17
What is a magnetic field?

A. Current flow through space around a permanent magnet
B. A force set up when current flows through a conductor
C. The force between the plates of a charged capacitor
D. The force that drives current through a resistor

3D18
In what direction is the magnetic field about a conductor when current is flowing?

A. In the same direction as the current
B. In a direction opposite to the current flow
C. In all directions; omnidirectional
D. In a direction determined by the left hand rule

3D19
What device is used to store electrical energy in an electrostatic field?

A. A battery
B. A transformer
C. A capacitor
D. An inductor

3D20
What is the term used to express the amount of electrical energy stored in an electrostatic field?

A. Coulombs
B. Joules
C. Watts
D. Volts

3D21
What factors determine the capacitance of a capacitor?

A. Area of the plates, voltage on the plates and distance between the plates
B. Area of the plates, distance between the plates and the dielectric constant of the material between the plates
C. Area of the plates, voltage on the plates and the dielectric constant of the material between the plates
D. Area of the plates, amount of charge on the plates and the dielectric constant of the material between the plates

3D22
What is the dielectric constant for air?

A. Approximately 1
B. Approximately 2
C. Approximately 4
D. Approximately 0

3D23
What determines the strength of the magnetic field around a conductor?

A. The resistance divided by the current
B. The ratio of the current to the resistance
C. The diameter of the conductor
D. The amount of current

3D24
Why would the rate at which electrical energy is used in a circuit be less than the product of the magnitudes of the AC voltage and current?

A. Because there is a phase angle that is greater than zero between the current and voltage
B. Because there are only resistances in the circuit
C. Because there are no reactances in the circuit
D. Because there is a phase angle that is equal to zero between the current and voltage

3D25
In a circuit where the AC voltage and current are out of phase, how can the true power be determined?

A. By multiplying the apparent power times the power factor
B. By subtracting the apparent power from the power factor
C. By dividing the apparent power by the power factor
D. By multiplying the RMS voltage times the RMS current

3D26
What does the power factor equal in an R-L circuit having a 60 degree phase angle between the voltage and the current?

A. 1.414
B. 0.866
C. 0.5
D. 1.73

3D27
What does the power factor equal in an R-L circuit having a 45 degree phase angle between the voltage and the current?

A. 0.866
B. 1.0
C. 0.5
D. 0.707

3D28
What does the power factor equal in an R-L circuit having a 30 degree phase angle between the voltage and the current?

A. 1.73
B. 0.5
C. 0.866
D. 0.577

3D29
How many watts are being consumed in a circuit having a power factor of 0.2 when the input is 100-VAC and 4-amperes is being drawn?

A. 400 watts
B. 80 watts
C. 2000 watts
D. 50 watts

3D30
How many watts are being consumed in a circuit having a power factor of 0.6 when the input is 200-VAC and 5-amperes is being drawn?

A. 200 watts
B. 1000 watts
C. 1600 watts
D. 600 watts

3D31
What is the effective radiated power of a repeater with 50 watts transmitter power output, 4 dB feedline loss, 3 dB duplexer and circulator loss, and 6 dB antenna gain?

A. 158 watts, assuming the antenna gain is referenced to a half-wave dipole
B. 39.7 watts, assuming the antenna gain is referenced to a half-wave dipole
C. 251 watts, assuming the antenna gain is referenced to a half-wave dipole
D. 69.9 watts, assuming the antenna gain is referenced to a half-wave dipole

3D32
What is the effective radiated power of a repeater with 50 watts transmitter power output, 5 dB feedline loss, 4 dB duplexer and circulator loss, and 7 dB antenna gain?

A. 300 watts, assuming the antenna gain is referenced to a half-wave dipole
B. 315 watts, assuming the antenna gain is referenced to a half-wave dipole
C. 31.5 watts, assuming the antenna gain is referenced to a half-wave dipole
D. 69.9 watts, assuming the antenna gain is referenced to a half-wave dipole

3D33
What is the effective radiated power of a repeater with 75 watts transmitter power output, 4 dB feedline loss, 3 dB duplexer and circulator loss, and 10 dB antenna gain?

A. 600 watts, assuming the antenna gain is referenced to a half-wave dipole
B. 75 watts, assuming the antenna gain is referenced to a half-wave dipole
C. 18.75 watts, assuming the antenna gain is referenced to a half-wave dipole
D. 150 watts, assuming the antenna gain is referenced to a half-wave dipole

3D34
What is the effective radiated power of a repeater with 75 watts transmitter power output, 5 dB feedline loss, 4 dB duplexer and circulator loss, and 6 dB antenna gain?

A. 37.6 watts, assuming the antenna gain is referenced to a half-wave dipole
B. 237 watts, assuming the antenna gain is referenced to a half-wave dipole
C. 150 watts, assuming the antenna gain is referenced to a half-wave dipole
D. 23.7 watts, assuming the antenna gain is referenced to a half-wave dipole

3D35
What is the effective radiated power of a repeater with 100 watts transmitter power output, 4 dB feedline loss, 3 dB duplexer and circulator loss, and 7 dB antenna gain?

A. 631 watts, assuming the antenna gain is referenced to a half-wave dipole
B. 400 watts, assuming the antenna gain is referenced to a half-wave dipole
C. 25 watts, assuming the antenna gain is referenced to a half-wave dipole
D. 100 watts, assuming the antenna gain is referenced to a half-wave dipole

3D36
What is the effective radiated power of a repeater with 100 watts transmitter power output, 5 dB feedline loss, 4 dB duplexer and circulator loss, and 10 dB antenna gain?

A. 800 watts, assuming the antenna gain is referenced to a half-wave dipole
B. 126 watts, assuming the antenna gain is referenced to a half-wave dipole
C. 12.5 watts, assuming the antenna gain is referenced to a half-wave dipole
D. 1260 watts, assuming the antenna gain is referenced to a half-wave dipole

3D37
What is the effective radiated power of a repeater with 120 watts transmitter power output, 5 dB feedline loss, 4 dB duplexer and circulator loss, and 6 dB antenna gain?

A. 601 watts, assuming the antenna gain is referenced to a half-wave dipole
B. 240 watts, assuming the antenna gain is referenced to a half-wave dipole
C. 60 watts, assuming the antenna gain is referenced to a half-wave dipole
D. 379 watts, assuming the antenna gain is referenced to a half-wave dipole

3D38
What is the effective radiated power of a repeater with 150 watts transmitter power output, 4 dB feedline loss, 3 dB duplexer and circulator loss, and 7 dB antenna gain?

A. 946 watts, assuming the antenna gain is referenced to a half-wave dipole
B. 37.5 watts, assuming the antenna gain is referenced to a half-wave dipole
C. 600 watts, assuming the antenna gain is referenced to a half-wave dipole
D. 150 watts, assuming the antenna gain is referenced to a half-wave dipole

3D39
What is the effective radiated power of a repeater with 200 watts transmitter power output, 4 dB feedline loss, 4 dB duplexer and circulator loss, and 10 dB antenna gain?

A. 317 watts, assuming the antenna gain is referenced to a half-wave dipole
B. 2000 watts, assuming the antenna gain is referenced to a half-wave dipole
C. 126 watts, assuming the antenna gain is referenced to a half-wave dipole
D. 260 watts, assuming the antenna gain is referenced to a half-wave dipole

3D40
What is the effective radiated power of a repeater with 200 watts transmitter power output, 4 dB feedline loss, 3 dB duplexer and circulator loss, and 6 dB antenna gain?

A. 252 watts, assuming the antenna gain is referenced to a half-wave dipole
B. 63.2 watts, assuming the antenna gain is referenced to a half-wave dipole
C. 632 watts, assuming the antenna gain is referenced to a half-wave dipole
D. 159 watts, assuming the antenna gain is referenced to a half-wave dipole

3D41
What is the photoconductive effect?

A. The conversion of photon energy to electromotive energy
B. The increased conductivity of an illuminated semiconductor junction
C. The conversion of electromotive energy to photon energy
D. The decreased conductivity of an illuminated semiconductor junction

3D42
What happens to photoconductive material when light shines on it?

A. The conductivity of the material increases
B. The conductivity of the material decreases
C. The conductivity of the material stays the same
D. The conductivity of the material becomes temperature dependent

3D43
What happens to the resistance of a photoconductive material when light shines on it?

A. It increases
B. It becomes temperature dependent
C. It stays the same
D. It decreases

3D44
What happens to the conductivity of a semiconductor junction when it is illuminated?

A. It stays the same
B. It becomes temperature dependent
C. It increases
D. It decreases

3D45
What is an optocoupler?

A. A resistor and a capacitor
B. A frequency modulated helium-neon laser
C. An amplitude modulated helium-neon laser
D. An LED and a phototransistor

3D46
What is an optoisolator?

A. An LED and a phototransistor
B. A P-N junction that develops an excess positive charge when exposed to light
C. An LED and a capacitor
D. An LED and a solar cell

3D47
What is an optical shaft encoder?

A. An array of optocouplers chopped by a stationary wheel
B. An array of optocouplers whose light transmission path is controlled by a rotating wheel
C. An array of optocouplers whose propagation velocity is controlled by a stationary wheel
D. An array of optocouplers whose propagation velocity is controlled by a rotating wheel

3D48
What does the photoconductive effect in crystalline solids produce a noticeable change in?

A. The capacitance of the solid
B. The inductance of the solid
C. The specific gravity of the solid
D. The resistance of the solid

3D49
What is the meaning of the term time constant of an RC circuit?

A. The time required to charge the capacitor in the circuit to 36.8% of the supply voltage
B. The time required to charge the capacitor in the circuit to 36.8% of the supply current
C. The time required to charge the capacitor in the circuit to 63.2% of the supply current
D. The time required to charge the capacitor in the circuit to 63.2% of the supply voltage

3D50
What is the meaning of the term time constant of an RL circuit?

A. The time required for the current in the circuit to build up to 36.8% of the maximum value
B. The time required for the voltage in the circuit to build up to 63.2% of the maximum value
C. The time required for the current in the circuit to build up to 63.2% of the maximum value
D. The time required for the voltage in the circuit to build up to 36.8% of the maximum value

3D51
What is the term for the time required for the capacitor in an RC circuit to be charged to 63.2% of the supply voltage?

A. An exponential rate of one
B. One time constant
C. One exponential period
D. A time factor of one

3D52
What is the term for the time required for the current in an RL circuit to build up to 63.2% of the maximum value?

A. One time constant
B. An exponential period of one
C. A time factor of one
D. One exponential rate

3D53
What is the term for the time it takes for a charged capacitor in an RC circuit to discharge to 36.8% of its initial value of stored charge?

A. One discharge period
B. An exponential discharge rate of one
C. A discharge factor of one
D. One time constant

3D54
What is meant by back EMF?

A. A current equal to the applied EMF
B. An opposing EMF equal to R times C (RC) percent of the applied EMF
C. A current that opposes the applied EMF
D. A voltage that opposes the applied EMF

3D55
After two time constants, the capacitor in an RC circuit is charged to what percentage of the supply voltage?

A. 36.8%
B. 63.2%
C. 86.5%
D. 95%

3D56
After two time constants, the capacitor in an RC circuit is discharged to what percentage of the starting voltage?

A. 86.5%
B. 63.2%
C. 36.8%
D. 13.5%

3D57
What is the time constant of a circuit having a 100-microfarad capacitor in series with a 470-kilohm resistor?

A. 4700 seconds
B. 470 seconds
C. 47 seconds
D. 0.47 seconds

3D58
What is the time constant of a circuit having a 220-microfarad capacitor in parallel with a 1-megohm resistor?

A. 220 seconds
B. 22 seconds
C. 2.2 seconds
D. 0.22 seconds

3D59
What is the time constant of a circuit having two 100-micrcfarad capacitors and two 470-kilohm resistors all in series?

A. 470 seconds
B. 47 seconds
C. 4.7 seconds
D. 0.47 seconds

3D60
What is the time constant of a circuit having two 100-micrcfarad capacitors and two 470-kilohm resistors all in parallel?

A. 470 seconds
B. 47 seconds
C. 4.7 seconds
D. 0.47 seconds

3D61
What is the time constant of a circuit having two 220-micrcfarad capacitors and two 1-megohm resistors all in series?

A. 55 seconds
B. 110 seconds
C. 220 seconds
D. 440 seconds

3D62
What is the time constant of a circuit having two 220-microfarad capacitors and two 1-megohm resistors all in parallel?

A. 22 seconds
B. 44 seconds
C. 220 seconds
D. 440 seconds

3D63
What is the time constant of a circuit having one 100-microfarad capacitor, one 220-microfarad capacitor, one 470-kilohm resistor and one 1-megohm resistor all in series?

A. 68.8 seconds
B. 101.1 seconds
C. 220.0 seconds
D. 470.0 seconds

3D64
What is the time constant of a circuit having a 470-microfarad capacitor and a 1-megohm resistor in parallel?

A. 0.47 seconds
B. 47 seconds
C. 220 seconds
D. 470 seconds

3D65
What is the time constant of a circuit having a 470-microfarad capacitor in series with a 470-kilohm resistor?

A. 221 seconds
B. 221,000 seconds
C. 470 seconds
D. 470,000 seconds

3D66
What is the time constant of a circuit having a 220-microfarad capacitor in series with a 470-kilohm resistor?

A. 103 seconds
B. 220 seconds
C. 470 seconds
D. 470,000 seconds

3D67
How long does it take for an initial charge of 20 V DC to decrease to 7.36 V DC in a 0.01-microfarad capacitor when a 2-megohm resistor is connected across it?

A. 12.64 seconds
B. 0.02 seconds
C. 1 second
D. 7.98 seconds

3D68
How long does it take for an initial charge of 20 V DC to decrease to 2.71 V DC in a 0.01-microfarad capacitor when a 2-megohm resistor is connected across it?

A. 0.04 seconds
B. 0.02 seconds
C. 7.36 seconds
D. 12.64 seconds

3D69
How long does it take for an initial charge of 20 V DC to decrease to 1 V DC in a 0.01-microfarad capacitor when a 2-megohm resistor is connected across it?

A. 0.01 seconds
B. 0.02 seconds
C. 0.04 seconds
D. 0.06 seconds

3D70
How long does it take for an initial charge of 20 V DC to decrease to 0.37 V DC in a 0.01-microfarad capacitor when a 2-megohm resistor is connected across it?

A. 0.08 seconds
B. 0.6 seconds
C. 0.4 seconds
D. 0.2 seconds

3D71
How long does it take for an initial charge of 20 V DC to decrease to 0.13 V DC in a 0.01-microfarad capacitor when a 2-megohm resistor is connected across it?

A. 0.06 seconds
B. 0.08 seconds
C. 0.1 seconds
D. 1.2 seconds

3D72
How long does it take for an initial charge of 800 V DC to decrease to 294 V DC in a 450-microfarad capacitor when a 1-megohm resistor is connected across it?

A. 80 seconds
B. 294 seconds
C. 368 seconds
D. 450 seconds

3D73
How long does it take for an initial charge of 800 V DC to decrease to 108 V DC in a 450-microfarad capacitor when a 1-megohm resistor is connected across it?

A. 225 seconds
B. 294 seconds
C. 450 seconds
D. 900 seconds

3D74
How long does it take for an initial charge of 800 V DC to decrease to 39.9 VDC in a 450-microfarad capacitor when a 1-megohm resistor is connected across it?

A. 1,350 seconds
B. 900 seconds
C. 450 seconds
D. 225 seconds

3D75
How long does it take for an initial charge of 800 V DC to decrease to 40.2 VDC in a 450-microfarad capacitor when a 1-megohm resistor is connected across it?

A. Approximately 225 seconds
B. Approximately 450 seconds
C. Approximately 900 seconds
D. Approximately 1,350 seconds

3D76
How long does it take for an initial charge of 800 V DC to decrease to 14.8 VDC in a 450-microfarad capacitor when a 1-megohm resistor is connected across it?

A. Approximately 900 seconds
B. Approximately 1,350 seconds
C. Approximately 1,804 seconds
D. Approximately 2,000 seconds

3D77
What is the impedance of a network comprised of a 0.1-microhenry inductor in series with a 20-ohm resistor, at 30 MHz? (Specify your answer in rectangular coordinates.)

A. 20 + j19
B. 20 - j19
C. 19 + j20
D. 19 - j20

3D78
What is the impedance of a network comprised of a 0.1-microhenry inductor in series with a 30-ohm resistor, at 5 MHz? (Specify your answer in rectangular coordinates.)

A. 30 - j3
B. 30 + j3
C. 3 + j30
D. 3 - j30

3D79
What is the impedance of a network comprised of a 10-microhenry inductor in series with a 40-ohm resistor, at 500 MHz? (Specify your answer in rectangular coordinates.)

A. 40 + j31400
B. 40 - j31400
C. 31400 + j40
D. 31400 - j40

3D80
What is the impedance of a network comprised of a 100-picofarad capacitor in parallel with a 4000-ohm resistor, at 500 kHz? (Specify your answer in polar coordinates.)

A. 2490 ohms, / 51.5 degrees
B. 4000 ohms, / 38.5 degrees
C. 5112 ohms, / -38.5 degrees
D. 2490 ohms, / -51.5 degrees

3D81
What is the impedance of a network comprised of a 0.001-microfarad capacitor in series with a 400-ohm resistor, at 500 kHz? (Specify your answer in rectangular coordinates.)

A. 400 - j318
B. 318 - j400
C. 400 + j318
D. 318 + j400

3D82
What is the impedance of a network comprised of a 100-ohm-reactance inductor in series with a 100-ohm resistor? (Specify your answer in polar coordinates.)

A. 121 ohms, / 35 degrees
B. 141 ohms, / 45 degrees
C. 161 ohms, / 55 degrees
D. 181 ohms, / 65 degrees

3D83
What is the impedance of a network comprised of a 100-ohm-reactance inductor, a 100-ohm-reactance capacitor, and a 100-ohm resistor all connected in series? (Specify your answer in polar coordinates.)

A. 100 ohms, / 90 degrees
B. 10 ohms, / 0 degrees
C. 100 ohms, / 0 degrees
D. 10 ohms, / 100 degrees

3D84
What is the impedance of a network comprised of a 400-ohm-reactance capacitor in series with a 300-ohm resistor? (Specify your answer in polar coordinates.)
A. 240 ohms, / 36.9 degrees
B. 240 ohms, / -36.9 degrees
C. 500 ohms, / 53.1 degrees
D. 500 ohms, / -53.1 degrees

3D85
What is the impedance of a network comprised of a 300-ohm-reactance capacitor, a 600-ohm-reactance inductor, and a 400-ohm resistor, all connected in series? (Specify your answer in polar coordinates.)

A. 500 ohms, / 37 degrees
B. 400 ohms, / 27 degrees
C. 300 ohms, / 17 degrees
D. 200 ohms, / 10 degrees

3D86
What is the impedance of a network comprised of a 400-ohm-reactance inductor in parallel with a 300-ohm resistor? (Specify your answer in polar coordinates.)

A. 240 ohms, / 36.9 degrees
B. 240 ohms, / -36.9 degrees
C. 500 ohms, / 53.1 degrees
D. 500 ohms, / -53.1 degrees

3D87
What is the impedance of a network comprised of a 1.0-millihenry inductor in series with a 200-ohm resistor, at 30 kHz? (Specify your answer in rectangular coordinates.)

A. 200 - j188
B. 200 + j188
C. 188 + j200
D. 188 - j200

3D88
What is the impedance of a network comprised of a 10-millihenry inductor in series with a 600-ohm resistor, at 10 kHz? (Specify your answer in rectangular coordinates.)

A. 628 + j600
B. 628 - j600
C. 600 + j628
D. 600 - j628

3D89
What is the impedance of a network comprised of a 0.01-microfarad capacitor in parallel with a 300-ohm resistor, at 50 kHz? (Specify your answer in rectangular coordinates.)

A. 150 - j159
B. 150 + j159
C. 159 + j150
D. 159 - j150

3D90
What is the impedance of a network comprised of a 0.1-microfarad capacitor in series with a 40-ohm resistor, at 50 kHz? (Specify your answer in rectangular coordinates.)

A. 40 + j32
B. 40 - j32
C. 32 - j40
D. 32 + j40

3D91
What is the impedance of a network comprised of a 1.0-microfarad capacitor in parallel with a 30-ohm resistor, at 5 MHz? (Specify your answer in rectangular coordinates.)

A. 0.000034 + j.032
B. 0.032 + j.000034
C. 0.000034 - j.032
D. 0.032 - j.000034

3D92
What is the impedance of a network comprised of a 100-ohm-reactance capacitor in series with a 100-ohm resistor? (Specify your answer in polar coordinates.)

A. 121 ohms, / -25 degrees
B. 141 ohms, / -45 degrees
C. 161 ohms, / -65 degrees
D. 191 ohms, / -85 degrees

3D93
What is the impedance of a network comprised of a 100-ohm-reactance capacitor in parallel with a 100-ohm resistor? (Specify your answer in polar coordinates.)

A. 31 ohms, / -15 degrees
B. 51 ohms, / -25 degrees
C. 71 ohms, / -45 degrees
D. 91 ohms, / -65 degrees

3D94
What is the impedance of a network comprised of a 300-ohm-reactance inductor in series with a 400-ohm resistor? (Specify your answer in polar coordinates.)

A. 400 ohms, / 27 degrees
B. 500 ohms, / 37 degrees
C. 600 ohms, / 47 degrees
D. 700 ohms, / 57 degrees

3D95
What is the impedance of a network comprised of a 100-ohm-reactance inductor in parallel with a 100-ohm resistor? (Specify your answer in polar coordinates.)
A. 71 ohms, / 45 degrees
B. 81 ohms, / 55 degrees
C. 91 ohms, / 65 degrees
D. 100 ohms, / 75 degrees

3D96
What is the impedance of a network comprised of a 300-ohm-reactance capacitor in series with a 400-ohm resistor? (Specify your answer in polar coordinates.)

A. 200 ohms, / -10 degrees
B. 300 ohms, / -17 degrees
C. 400 ohms, / -27 degrees
D. 500 ohms, / -37 degrees

3D97
The speed of a series DC motor will be affected by the:

A. line voltage frequency
B. number of windings
C. load
D. rotor coil windings

3D98
The expression 'voltage regulation' as it applies to a shunt-wound DC generator operating at a constant frequency refers to:

A. voltage output efficiency
B. voltage in the secondary compared to the primary
C. voltage fluctuations from load to no-load
D. rotor winding voltage ratio

3D99
When an emergency transmitter uses 325 watts and a receiver uses 50 watts, how many hours can a 12.6 volt, 55 ampere-hour battery supply full power to both units?

A. 6 hours
B. 3 hours
C. 1.8 hours
D. 1.2 hours

3D100
The expression "voltage regulation" as it applies to a generator operating at a constant frequency refers to:

A. full load to no load
B. limited load to peak load
C. source input supply frequency
D. field frequency

3D101
The output of a separately excited AC generator running at a constant speed can be controlled by:

A. armature
B. brushes
C. field current
D. exciter

3D102
What occurs if the load is removed from an operating series DC motor?

A. it will stop running
B. speed will increase slightly
C. no change occurs
D. it will accelerate until it flies apart

3D103
The speed of a DC motor varies with the:

A. load
B. number of brushes
C. ripple frequency
D. RF field voltage

3D104
A 12.6 volt, 8 ampere-hour battery is supplying power to a receiver which uses 50 watts and a radar system that uses 300 watts. How long will the battery last?

A. 100.8 hours
B. 27.7 hours
C. 1 hour
D. 17 minutes or 0.3 hours

3D105
What is the total voltage when 12 Nickel-Cadmium batteries are connected in series?

A. 12 volts
B. 12.6 volts
C. 15 volts
D. 72 volts

3D106
A ship radar unit uses 315 watts and a radio uses 50 watts. If the equipment is connected to a 50 ampere-hour battery rated at 12.6 volts, how long will the battery last?

A. 28.97 hours
B. 29 minutes
C. 1 hour 43 minutes
D. 10 hours 50 minutes

3D107
The output voltage of a separately excited AC generator (running at a constant speed) is controlled by:

A. input voltage frequency
B. the load
C. primary voltage
D. field current

3D108
There is an improper impedance match between a 30 watt transmitter and the antenna and 5 watts is reflected. How much power is actually radiated?

A. 35 watts
B. 30 watts
C. 25 watts
D. 20 watts

3D109
How long will a 12.6 volt, 50 ampere-hour battery last if it supplies power to an emergency transmitter rated at 531 watts of plate input power and other emergency equipment with a combined power rating of 530 watts?

A. 6 hours
B. 4 hours
C. 1 hour
D. 35 minutes

3D110
A 12.6 volt, 55 ampere-hour battery is connected to a radar unit rated at 325 watts and a receiver that uses 20 watts. How long will radar unit and receiver be able to draw full power from the battery?

A. 6 hours
B. 4 hours
C. 2.3 hours
D. 2 hours

3D111
The speed of a series wound DC motor varies with:

A. the number of commutator bars
B. the number of slip rings
C. the load applied to the motor
D. the direction of rotation

3D112
A 6 volt battery with 1.2 ohms internal resistance is connected across two 3 watt bulbs. What is the current flow?

A. .57 amps
B. .83 amps
C. 1.0 amps
D. 6.0 amps

3D113
A dynamotor is approximately:

A. 100% efficient
B. 85% efficient
C. 65% efficient
D. 40% efficient

3D114
The power input to a 52 ohm transmission line is 1,872 watts. The current flowing through the line is:

A. 6 amps
B. 144 amps
C. 0.06 amps
D. 28.7 amps

3D115
If a marine radiotelephone receiver uses 75 watts of power and a transmitter uses 325 watts, how long can they both operate before discharging a 50 ampere-hour 12 volt battery?

A. 40 minutes
B. 1 hour
C. 1 1/2 hours
D. 6 hours

Test 3-E Element Three
Circuit Components

3E1
Structurally, what are the two main categories of semiconductor diodes?

A. Junction and point contact
B. Electrolytic and junction
C. Electrolytic and point contact
D. Vacuum and point contact

3E2
What are the two primary classifications of Zener diodes?

A. Hot carrier and tunnel
B. Varactor and rectifying
C. Voltage regulator and voltage reference
D. Forward and reversed biased

3E3
What is the principal characteristic of a Zener diode?

A. A constant current under conditions of varying voltage
B. A constant voltage under conditions of varying current
C. A negative resistance region
D. An internal capacitance that varies with the applied voltage

3E4
What is the range of voltage ratings available in Zener diodes?

A. 2.4 volts to 200 volts
B. 1.2 volts to 7 volts
C. 3 volts to 2000 volts
D. 1.2 volts to 5.6 volts

3E5
What is the principal characteristic of a tunnel diode?

A. A high forward resistance
B. A very high PIV
C. A negative resistance region
D. A high forward current rating

3E6
What special type of diode is capable of both amplification and oscillation?

A. Point contact diodes
B. Zener diodes
C. Tunnel diodes
D. Junction diodes

3E7
What type of semiconductor diode varies its internal capacitance as the voltage applied to its terminals varies?

A. A varactor diode
B. A tunnel diode
C. A silicon-controlled rectifier
D. A Zener diode

3E8
What is the principal characteristic of a varactor diode?

A. It has a constant voltage under conditions of varying current
B. Its internal capacitance varies with the applied voltage
C. It has a negative resistance region
D. It has a very high PIV

3E9
What is a common use of a varactor diode?

A. As a constant current source
B. As a constant voltage source
C. As a voltage controlled inductance
D. As a voltage controlled capacitance

3E10
What is a common use of a hot-carrier diode?

A. As balanced mixers in SSB generation
B. As a variable capacitance in an automatic frequency control circuit
C. As a constant voltage reference in a power supply
D. As VHF and UHF mixers and detectors

3E11
What limits the maximum forward current in a junction diode?

A. The peak inverse voltage
B. The junction temperature
C. The forward voltage
D. The back EMF

3E12
How are junction diodes rated?

A. Maximum forward current and capacitance
B. Maximum reverse current and PIV
C. Maximum reverse current and capacitance
D. Maximum forward current and PIV

3E13
What is a common use for point contact diodes?

A. As a constant current source
B. As a constant voltage source
C. As an RF detector
D. As a high voltage rectifier

3E14
What type of diode is made of a metal whisker touching a very small semiconductor die?

A. Zener diode
B. Varactor diode
C. Junction diode
D. Point contact diode

3E15
What is one common use for PIN diodes?

A. As a constant current source
B. As a constant voltage source
C. As an RF switch
D. As a high voltage rectifier

3E16
What special type of diode is often used in RF switches, attenuators, and various types of phase shifting devices?

A. Tunnel diodes
B. Varactor diodes
C. PIN diodes
D. Junction diodes

3E17
What are the three terminals of a bipolar transistor?

A. Cathode, plate and grid
B. Base, collector and emitter
C. Gate, source and sink
D. Input, output and ground

3E18
What is the meaning of the term alpha with regard to bipolar transistors?

A. The change of collector current with respect to base current
B. The change of base current with respect to collector current
C. The change of collector current with respect to emitter current
D. The change of collector current with respect to gate current

3E19
What is the term used to express the ratio of change in DC collector current to a change in emitter current in a bipolar transistor?

A. Gamma
B. Epsilon
C. Alpha
D. Beta

3E20
What is the meaning of the term beta with regard to bipolar transistors?

A. The change of collector current with respect to base current
B. The change of base current with respect to emitter current
C. The change of collector current with respect to emitter current
D. The change in base current with respect to gate current

3E21
What is the term used to express the ratio of change in the DC collector current to a change in base current in a bipolar transistor?

A. Alpha
B. Beta
C. Gamma
D. Delta

3E22
What is the meaning of the term alpha cutoff frequency with regard to bipolar transistors?

A. The practical lower frequency limit of a transistor in common emitter configuration
B. The practical upper frequency limit of a transistor in common base configuration
C. The practical lower frequency limit of a transistor in common base configuration
D. The practical upper frequency limit of a transistor in common emitter configuration

3E23
What is the term used to express that frequency at which the grounded base current gain has decreased to 0.7 of the gain obtainable at 1 kHz in a transistor?

A. Corner frequency
B. Alpha cutoff frequency
C. Beta cutoff frequency
D. Alpha rejection frequency

3E24
What is the meaning of the term beta cutoff frequency with regard to a bipolar transistor?

A. That frequency at which the grounded base current gain has decreased to 0.7 of that obtainable at 1 kHz in a transistor
B. That frequency at which the grounded emitter current gain has decreased to 0.7 of that obtainable at 1 kHz in a transistor
C. That frequency at which the grounded collector current gain has decreased to 0.7 of that obtainable at 1 kHz in a transistor
D. That frequency at which the grounded gate current gain has decreased to 0.7 of that obtainable at 1 kHz in a transistor

3E25
What is the meaning of the term transition region with regard to a transistor?

A. An area of low charge density around the P-N junction
B. The area of maximum P-type charge
C. The area of maximum N-type charge
D. The point where wire leads are connected to the P- or N-type material

3E26
What does it mean for a transistor to be fully saturated?

A. The collector current is at its maximum value
B. The collector current is at its minimum value
C. The transistor's Alpha is at its maximum value
D. The transistor's Beta is at its maximum value

3E27
What does it mean for a transistor to be cut off?

A. There is no base current
B. The transistor is at its operating point
C. No current flows from emitter to collector
D. Maximum current flows from emitter to collector

3E28
What are the elements of a unijunction transistor?

A. Base 1, base 2 and emitter
B. Gate, cathode and anode
C. Gate, base 1 and base 2
D. Gate, source and sink

3E29
For best efficiency and stability, where on the load-line should a solid-state power amplifier be operated?

A. Just below the saturation point
B. Just above the saturation point
C. At the saturation point
D. At 1.414 times the saturation point

3E30
What two elements widely used in semiconductor devices exhibit both metallic and non-metallic characteristics?

A. Silicon and gold
B. Silicon and germanium
C. Galena and germanium
D. Galena and bismuth

3E31
What are the three terminals of an SCR?

A. Anode, cathode and gate
B. Gate, source and sink
C. Base, collector and emitter
D. Gate, base 1 and base 2

3E32
What are the two stable operating conditions of an SCR?

A. Conducting and nonconducting
B. Oscillating and quiescent
C. Forward conducting and reverse conducting
D. NPN conduction and PNP conduction

3E33
When an SCR is in the triggered or on condition, its electrical characteristics are similar to what other solid-state device (as measured between its cathode and anode)?

A. The junction diode
B. The tunnel diode
C. The hot-carrier diode
D. The varactor diode

3E34
Under what operating condition does an SCR exhibit electrical characteristics similar to a forward-biased silicon rectifier?

A. During a switching transition
B. When it is used as a detector
C. When it is gated "off"
D. When it is gated "on"

3E35
What is the transistor called which is fabricated as two complementary SCRs in parallel with a common gate terminal?

A. TRIAC
B. Bilateral SCR
C. Unijunction transistor
D. Field effect transistor

3E36
What are the three terminals of a TRIAC?

A. Emitter, base 1 and base 2
B. Gate, anode 1 and anode 2
C. Base, emitter and collector
D. Gate, source and sink

3E37
What is the normal operating voltage and current for a light-emitting diode?

A. 60 volts and 20 mA
B. 5 volts and 50 mA
C. 1.7 volts and 20 mA
D. 0.7 volts and 60 mA

3E38
What type of bias is required for an LED to produce luminescence?

A. Reverse bias
B. Forward bias
C. Zero bias
D. Inductive bias

3E39
What are the advantages of using an LED?

A. Low power consumption and long life
B. High lumens per cm per cm and low power consumption
C. High lumens per cm per cm and low voltage requirement
D. A current flows when the device is exposed to a light source

3E40
What colors are available in LEDs?

A. Yellow, blue, red and brown
B. Red, violet, yellow and peach
C. Violet, blue, orange and red
D. Red, green, orange and yellow

3E41
How can a neon lamp be used to check for the presence of RF?

A. A neon lamp will go out in the presence of RF
B. A neon lamp will change color in the presence of RF
C. A neon lamp will light only in the presence of very low frequency RF
D. A neon lamp will light in the presence of RF

3E42
What would be the bandwidth of a good crystal lattice band-pass filter for a single-sideband phone emission?

A. 6 kHz at -6 dB
B. 2.1 kHz at -6 dB
C. 500 Hz at -6 dB
D. 15 kHz at -6 dB

3E43
What would be the bandwidth of a good crystal lattice band-pass filter for a double-sideband phone emission?

A. 1 kHz at -6 dB
B. 500 Hz at -6 dB
C. 6 kHz at -6 dB
D. 15 kHz at -6 dB

3E44
What is a crystal lattice filter?

A. A power supply filter made with crisscrossed quartz crystals
B. An audio filter made with 4 quartz crystals at 1-kHz intervals
C. A filter with infinitely wide and shallow skirts made using quartz crystals
D. A filter with narrow bandwidth and steep skirts made using quartz crystals

3E45
What technique can be used to construct low cost, high performance crystal lattice filters?

A. Splitting and tumbling
B. Tumbling and grinding
C. Etching and splitting
D. Etching and grinding

3E46
What determines the bandwidth and response shape in a crystal lattice filter?

A. The relative frequencies of the individual crystals
B. The center frequency chosen for the filter
C. The amplitude of the RF stage preceding the filter
D. The amplitude of the signals passing through the filter

3E47
What is an enhancement-mode FET?

A. An FET with a channel that blocks voltage through the gate
B. An FET with a channel that allows a current when the gate voltage is zero
C. An FET without a channel to hinder current through the gate
D. An FET without a channel; no current occurs with zero gate voltage

3E48
What is a depletion-mode FET?

A. An FET that has a channel with no gate voltage applied; a current flows with zero gate voltage
B. An FET that has a channel that blocks current when the gate voltage is zero
C. An FET without a channel; no current flows with zero gate voltage
D. An FET without a channel to hinder current through the gate

3E49
Why do many MOSFET devices have built-in gate-protective Zener diodes?

A. The gate-protective Zener diode provides a voltage reference to provide the correct amount of reverse-bias gate voltage
B. The gate-protective Zener diode protects the substrate from excessive voltages
C. The gate-protective Zener diode keeps the gate voltage within specifications to prevent the device from overheating
D. The gate-protective Zener diode prevents the gate insulation from being punctured by small static charges or excessive voltages

3E50
What do the initials CMOS stand for?

A. Common mode oscillating system
B. Complementary mica-oxide silicon
C. Complementary metal-oxide semiconductor
D. Complementary metal-oxide substrate

3E51
Why are special precautions necessary in handling FET and CMOS devices?

A. They are susceptible to damage from static charges
B. They have fragile leads that may break off
C. They have micro-welded semiconductor junctions that are susceptible to breakage
D. They are light sensitive

3E52
How does the input impedance of a field-effect transistor compare with that of a bipolar transistor?

A. One cannot compare input impedance without first knowing the supply voltage
B. An FET has low input impedance; a bipolar transistor has high input impedance
C. The input impedance of FETs and bipolar transistors is the same
D. An FET has high input impedance; a bipolar transistor has low input impedance

3E53
What are the three terminals of a field-effect transistor?

A. Gate 1, gate 2, drain
B. Emitter, base, collector
C. Emitter, base 1, base 2
D. Gate, drain, source

3E54
What are the two basic types of junction field-effect transistors?

A. N-channel and P-channel
B. High power and low power
C. MOSFET and GaAsFET
D. Silicon FET and germanium FET

3E55
What is an operational amplifier?

A. A high-gain, direct-coupled differential amplifier whose characteristics are determined by components external to the amplifier unit
B. A high-gain, direct-coupled audio amplifier whose characteristics are determined by components external to the amplifier unit
C. An amplifier used to increase the average output of frequency modulated signals
D. A program subroutine that calculates the gain of an RF amplifier

3E56
What would be the characteristics of the ideal op-amp?

A. Zero input impedance, infinite output impedance, infinite gain, flat frequency response
B. Infinite input impedance, zero output impedance, infinite gain, flat frequency response
C. Zero input impedance, zero output impedance, infinite gain, flat frequency response
D. Infinite input impedance, infinite output impedance, infinite gain, flat frequency response

3E57
What determines the gain of a closed-loop op-amp circuit?

A. The external feedback network
B. The collector-to-base capacitance of the PNP stage
C. The power supply voltage
D. The PNP collector load

3E58
What is meant by the term op-amp offset voltage?

A. The output voltage of the op-amp minus its input voltage
B. The difference between the output voltage of the op-amp and the input voltage required in the following stage
C. The potential between the amplifier-input terminals of the op-amp in a closed-loop condition
D. The potential between the amplifier-input terminals of the op-amp in an open-loop condition

3E59
What is the input impedance of a theoretically ideal op-amp?

A. 100 ohms
B. 1,000 ohms
C. Very low
D. Very high

3E60
What is the output impedance of a theoretically ideal op-amp?

A. Very low
B. Very high
C. 100 ohms
D. 1000 ohms

3E61
What is a phase-locked loop circuit?

A. An electronic servo loop consisting of a ratio detector, reactance modulator,and voltage-controlled oscillator
B. An electronic circuit also known as a monostable multivibrator
C. An electronic circuit consisting of a precision push-pull amplifier with a differential input
D. An electronic servo loop consisting of a phase detector, a low-pass filter and voltage-controlled oscillator

3E62
What functions are performed by a phase-locked loop?

A. Wideband AF and RF power amplification
B. Comparison of two digital input signals, digital pulse counter
C. Photovoltaic conversion, optical coupling
D. Frequency synthesis, FM demodulation

3E63
A circuit compares the output from a voltage-controlled oscillator and a frequency standard. The difference between the two frequencies produces an error voltage that changes the voltage-controlled oscillator frequency. What is the name of the circuit?

A. A doubly balanced mixer
B. A phase-locked loop
C. A differential voltage amplifier
D. A variable frequency oscillator

3E64
What do the initials TTL stand for?

A. Resistor-transistor logic
B. Transistor-transistor logic
C. Diode-transistor logic
D. Emitter-coupled logic

3E65
What is the recommended power supply voltage for TTL series integrated circuits?

A. 12.00 volts
B. 50.00 volts
C. 5.00 volts
D. 13.60 volts

3E66
What logic state do the inputs of a TTL device assume if they are left open?

A. A high logic state
B. A low logic state
C. The device becomes randomized and will not provide consistent high or low logic states
D. Open inputs on a TTL device are ignored

3E67
What level of input voltage is high in a TTL device operating with a 5-volt power supply?

A. 2.0 to 5.5 volts
B. 1.5 to 3.0 volts
C. 1.0 to 1.5 volts
D. -5.0 to -2.0 volts

3E68
What level of input voltage is low in a TTL device operating with a 5-volt power supply?

A. -2.0 to -5.5 volts
B. 2.0 to 5.5 volts
C. -0.6 to 0.8 volts
D. -0.8 to 0.4 volts

3E69
Why do circuits containing TTL devices have several bypass capacitors per printed circuit board?

A. To prevent RFI to receivers
B. To keep the switching noise within the circuit, thus eliminating RFI
C. To filter out switching harmonics
D. To prevent switching transients from appearing on the supply line

3E70
What is a CMOS IC?

A. A chip with only P-channel transistors
B. A chip with P-channel and N-channel transistors
C. A chip with only N-channel transistors
D. A chip with only bipolar transistors

3E71
What is one major advantage of CMOS over other devices?

A. Small size
B. Low current consumption
C. Low cost
D. Ease of circuit design

3E72
Why do CMOS digital integrated circuits have high immunity to noise on the input signal or power supply?

A. Larger bypass capacitors are used in CMOS circuit design
B. The input switching threshold is about two times the power supply voltage
C. The input switching threshold is about one-half the power supply voltage
D. Input signals are stronger

3E73
Signal energy is coupled into a traveling-wave tube at:

A. collector end of helix
B. anode end of the helix
C. cathode end of the helix
D. focusing coils

3E74
Permanent magnetic field that surrounds a traveling-wave tube (TWT) is intended to:

A. provide a means of coupling
B. prevent the electron beam from spreading
C. prevent oscillations
D. prevent spurious oscillations

3E75
Electromagnetic coils encase a traveling wave tube to:

A. provide a means of coupling energy
B. prevent the electron beam from spreading
C. prevent oscillation
D. prevent spurious oscillation

Test 3-F Element Three
Practical Circuits

3F1
What is a linear electronic voltage regulator?

A. A regulator that has a ramp voltage as its output
B. A regulator in which the pass transistor switches from the "off" state to the "on" state
C. A regulator in which the control device is switched on or off, with the duty cycle proportional to the line or load conditions
D. A regulator in which the conduction of a control element is varied in directproportion to the line voltage or load current

3F2
What is a switching electronic voltage regulator?

A. A regulator in which the conduction of a control element is varied in direct proportion to the line voltage or load current
B. A regulator that provides more than one output voltage
C. A regulator in which the control device is switched on or off, with the duty cycle proportional to the line or load conditions
D. A regulator that gives a ramp voltage at its output

3F3
What device is usually used as a stable reference voltage in a linear voltage regulator?

A. A Zener diode
B. A tunnel diode
C. An SCR
D. A varactor diode

3F4
What type of linear regulator is used in applications requiring efficient utilization of the primary power source?

A. A constant current source
B. A series regulator
C. A shunt regulator
D. A shunt current source

3F5
What type of linear voltage regulator is used in applications where the load on the unregulated voltage source must be kept constant?

A. A constant current source
B. A series regulator
C. A shunt current source
D. A shunt regulator

3F6
To obtain the best temperature stability, what should be the operating voltage of the reference diode in a linear voltage regulator?

A. Approximately 2.0 volts
B. Approximately 3.0 volts
C. Approximately 6.0 volts
D. Approximately 10.0 volts

3F7
What is the meaning of the term remote sensing with regard to a linear voltage regulator?

A. The feedback connection to the error amplifier is made directly to the load
B. Sensing is accomplished by wireless inductive loops
C. The load connection is made outside the feedback loop
D. The error amplifier compares the input voltage to the reference voltage

3F8
What is a three-terminal regulator?

A. A regulator that supplies three voltages with variable current
B. A regulator that supplies three voltages at a constant current
C. A regulator containing three error amplifiers and sensing transistors
D. A regulator containing a voltage reference, error amplifier, sensing resistors and transistors, and a pass element

3F9
What are the important characteristics of a three-terminal regulator?

A. Maximum and minimum input voltage, minimum output current and voltage
B. Maximum and minimum input voltage, maximum output current and voltage
C. Maximum and minimum input voltage, minimum output current and maximum output voltage
D. Maximum and minimum input voltage, minimum output voltage and maximum output current

3F10
What is the distinguishing feature of a Class A amplifier?

A. Output for less than 180 degrees of the signal cycle
B. Output for the entire 360 degrees of the signal cycle
C. Output for more than 180 degrees and less than 360 degrees of the signal cycle
D. Output for exactly 180 degrees of the input signal cycle

3F11
What class of amplifier is distinguished by the presence of output throughout the entire signal cycle and the input never goes into the cutoff region?

A. Class A
B. Class B
C. Class C
D. Class D

3F12
What is the distinguishing characteristic of a Class B amplifier?

A. Output for the entire input signal cycle
B. Output for greater than 180 degrees and less than 360 degrees of the input signal cycle
C. Output for less than 180 degrees of the input signal cycle
D. Output for 180 degrees of the input signal cycle

3F13
What class of amplifier is distinguished by the flow of current in the output essentially in 180 degree pulses?

A. Class A
B. Class B
C. Class C
D. Class D

3F14
What is a Class AB amplifier?

A. Output is present for more than 180 degrees but less than 360 degrees of the signal input cycle
B. Output is present for exactly 180 degrees of the input signal cycle
C. Output is present for the entire input signal cycle
D. Output is present for less than 180 degrees of the input signal cycle

3F15
What is the distinguishing feature of a Class C amplifier?

A. Output is present for less than 180 degrees of the input signal cycle
B. Output is present for exactly 180 degrees of the input signal cycle
C. Output is present for the entire input signal cycle
D. Output is present for more than 180 degrees but less than 360 degrees of the input signal cycle

3F16
What class of amplifier is distinguished by the bias being set well beyond cutoff?

A. Class A
B. Class B
C. Class C
D. Class AB

3F17
Which class of amplifier provides the highest efficiency?

A. Class A
B. Class B
C. Class C
D. Class AB

3F18
Which class of amplifier has the highest linearity and least distortion?

A. Class A
B. Class B
C. Class C
D. Class AB

3F19
Which class of amplifier has an operating angle of more than 180 degrees but less than 360 degrees when driven by a sine wave signal?

A. Class A
B. Class B
C. Class C
D. Class AB

3F20
What is an L-network?

A. A network consisting entirely of four inductors
B. A network consisting of an inductor and a capacitor
C. A network used to generate a leading phase angle
D. A network used to generate a lagging phase angle

3F21
What is a pi-network?

A. A network consisting entirely of four inductors or four capacitors
B. A Power Incidence network
C. An antenna matching network that is isolated from ground
D. A network consisting of one inductor and two capacitors or two inductors and one capacitor

3F22
What is a pi-L-network?

A. A Phase Inverter Load network
B. A network consisting of two inductors and two capacitors
C. A network with only three discrete parts
D. A matching network in which all components are isolated from ground

3F23
Does the L-, pi-, or pi-L-network provide the greatest harmonic suppression?

A. L-network
B. Pi-network
C. Inverse L-network
D. Pi-L-network

3F24
What are the three most commonly used networks to accomplish a match between an amplifying device and a transmission line?

A. M-network, pi-network and T-network
B. T-network, M-network and Q-network
C. L-network, pi-network and pi-L-network
D. L-network, M-network and C-network

3F25
How are networks able to transform one impedance to another?

A. Resistances in the networks substitute for resistances in the load
B. The matching network introduces negative resistance to cancel the resistive part of an impedance
C. The matching network introduces transconductance to cancel the reactive part of an impedance
D. The matching network can cancel the reactive part of an impedance and change the value of the resistive part of an impedance

3F26
Which type of network offers the greater transformation ratio?

A. L-network
B. Pi-network
C. Constant-K
D. Constant-M

3F27
Why is the L-network of limited utility in impedance matching?

A. It matches a small impedance range
B. It has limited power handling capabilities
C. It is thermally unstable
D. It is prone to self resonance

3F28
What is an advantage of using a pi-L-network instead of a pi-network for impedance matching between the final amplifier of a vacuum-tube type transmitter and a multiband antenna?

A. Greater transformation range
B. Higher efficiency
C. Lower losses
D. Greater harmonic suppression

3F29
Which type of network provides the greatest harmonic suppression?

A. L-network
B. Pi-network
C. Pi-L-network
D. Inverse-Pi network

3F30
What are the three general groupings of filters?

A. High-pass, low-pass and band-pass
B. Inductive, capacitive and resistive
C. Audio, radio and capacitive
D. Hartley, Colpitts and Pierce

3F31
What is a constant-K filter?

A. A filter that uses Boltzmann's constant
B. A filter whose velocity factor is constant over a wide range of frequencies
C. A filter whose product of the series- and shunt-element impedances is a constant for all frequencies
D. A filter whose input impedance varies widely over the design bandwidth

3F32
What is an advantage of a constant-k filter?

A. It has high attenuation for signals on frequencies far removed from the passband
B. It can match impedances over a wide range of frequencies
C. It uses elliptic functions
D. The ratio of the cutoff frequency to the trap frequency can be varied

3F33
What is an m-derived filter?

A. A filter whose input impedance varies widely over the design bandwidth
B. A filter whose product of the series- and shunt-element impedances is a constant for all frequencies
C. A filter whose schematic shape is the letter "M"
D. A filter that uses a trap to attenuate undesired frequencies too near cutoff for a constant-k filter.

3F34
What are the distinguishing features of a Butterworth filter?

A. A filter whose product of the series- and shunt-element impedances is a constant for all frequencies
B. It only requires capacitors
C. It has a maximally flat response over its passband
D. It requires only inductors

3F35
What are the distinguishing features of a Chebyshev filter?

A. It has a maximally flat response over its passband
B. It allows ripple in the passband
C. It only requires inductors
D. A filter whose product of the series- and shunt-element impedances is a constant for all frequencies

3F36
When would it be more desirable to use an m-derived filter over a constant-k filter?

A. When the response must be maximally flat at one frequency
B. When you need more attenuation at a certain frequency that is too close to the cut-off frequency for a constant-k filter
C. When the number of components must be minimized
D. When high power levels must be filtered

3F37
What are three major oscillator circuits often used in radio equipment?

A. Taft, Pierce and negative feedback
B. Colpitts, Hartley and Taft
C. Taft, Hartley and Pierce
D. Colpitts, Hartley and Pierce

3F38
How is the positive feedback coupled to the input in a Hartley oscillator?

A. Through a neutralizing capacitor
B. Through a capacitive divider
C. Through link coupling
D. Through a tapped coil

3F39
How is the positive feedback coupled to the input in a Colpitts oscillator?

A. Through a tapped coil
B. Through link coupling
C. Through a capacitive divider
D. Through a neutralizing capacitor

3F40
How is the positive feedback coupled to the input in a Pierce oscillator?

A. Through a tapped coil
B. Through link coupling
C. Through a capacitive divider
D. Through capacitive coupling

3F41
Which of the three major oscillator circuits used in radio equipment utilizes a quartz crystal?

A. Negative feedback
B. Hartley
C. Colpitts
D. Pierce

3F42
What is the piezoelectric effect?

A. Mechanical vibration of a crystal by the application of a voltage
B. Mechanical deformation of a crystal by the application of a magnetic field
C. The generation of electrical energy by the application of light
D. Reversed conduction states when a P-N junction is exposed to light

3F43
What is the major advantage of a Pierce oscillator?

A. It is easy to neutralize
B. It doesn't require an LC tank circuit
C. It can be tuned over a wide range
D. It has a high output power

3F44
Which type of oscillator circuit is commonly used in a VFO?

A. Pierce
B. Colpitts
C. Hartley
D. Negative feedback

3F45
Why is the Colpitts oscillator circuit commonly used in a VFO?

A. The frequency is a linear function of the load impedance
B. It can be used with or without crystal lock-in
C. It is stable
D. It has high output power

3F46
What is meant by the term modulation?

A. The squelching of a signal until a critical signal-to-noise ratio is reached
B. Carrier rejection through phase nulling
C. A linear amplification mode
D. A mixing process whereby information is imposed upon a carrier

3F47
How is an G3E FM-phone emission produced?

A. With a balanced modulator on the audio amplifier
B. With a reactance modulator on the oscillator
C. With a reactance modulator on the final amplifier
D. With a balanced modulator on the oscillator

3F48
What is a reactance modulator?

A. A circuit that acts as a variable resistance or capacitance to produce FM signals
B. A circuit that acts as a variable resistance or capacitance to produce AM signals
C. A circuit that acts as a variable inductance or capacitance to produce FM signals
D. A circuit that acts as a variable inductance or capacitance to produce AM signals

3F49
What is a balanced modulator?

A. An FM modulator that produces a balanced deviation
B. A modulator that produces a double sideband, suppressed carrier signal
C. A modulator that produces a single sideband, suppressed carrier signal
D. A modulator that produces a full carrier signal

3F50
How can a single-sideband phone signal be generated?

A. By driving a product detector with a DSB signal
B. By using a reactance modulator followed by a mixer
C. By using a loop modulator followed by a mixer
D. By using a balanced modulator followed by a filter

3F51
How can a double-sideband phone signal be generated?

A. By feeding a phase modulated signal into a low pass filter
B. By using a balanced modulator followed by a filter
C. By detuning a Hartley oscillator
D. By modulating the plate voltage of a class C amplifier

3F52
How is the efficiency of a power amplifier determined?

A. Efficiency = (RF power out / DC power in) X 100%
B. Efficiency = (RF power in / RF power out) X 100%
C. Efficiency = (RF power in / DC power in) X 100%
D. Efficiency = (DC power in / RF power in) X 100%

3F53
For reasonably efficient operation of a transistor amplifier, what should the load resistance be with 12 volts at the collector and 5 watts power output?

A. 100.3 ohms
B. 14.4 ohms
C. 10.3 ohms
D. 144 ohms

3F54
What is the flywheel effect?

A. The continued motion of a radio wave through space when the transmitter is turned off
B. The back and forth oscillation of electrons in an LC circuit
C. The use of a capacitor in a power supply to filter rectified AC
D. The transmission of a radio signal to a distant station by several hops through the ionosphere

3F55
What order of Q is required by a tank-circuit sufficient to reduce harmonics to an acceptable level?

A. Approximately 120
B. Approximately 12
C. Approximately 1200
D. Approximately 1.2

3F56
How can parasitic oscillations be eliminated from a power amplifier?

A. By tuning for maximum SWR
B. By tuning for maximum power output
C. By neutralization
D. By tuning the output

3F57
What is the process of detection?

A. The process of masking out the intelligence on a received carrier to make an S-meter operational
B. The recovery of intelligence from the modulated RF signal
C. The modulation of a carrier
D. The mixing of noise with the received signal

3F58
What is the principle of detection in a diode detector?

A. Rectification and filtering of RF
B. Breakdown of the Zener voltage
C. Mixing with noise in the transition region of the diode
D. The change of reactance in the diode with respect to frequency

3F59
What is a product detector?

A. A detector that provides local oscillations for input to the mixer
B. A detector that amplifies and narrows the band-pass frequencies
C. A detector that uses a mixing process with a locally generated carrier
D. A detector used to detect cross-modulation products

3F60
How are FM-phone signals detected?

A. By a balanced modulator
B. By a frequency discriminator
C. By a product detector
D. By a phase splitter

3F61
What is a frequency discriminator?

A. A circuit for detecting FM signals
B. A circuit for filtering two closely adjacent signals
C. An automatic bandswitching circuit
D. An FM generator

3F62
What is the mixing process?

A. The elimination of noise in a wideband receiver by phase comparison
B. The elimination of noise in a wideband receiver by phase differentiation
C. Distortion caused by auroral propagation
D. The combination of two signals to produce sum and difference frequencies

3F63
What are the principal frequencies which appear at the output of a mixer circuit?

A. Two and four times the original frequency
B. The sum, difference and square root of the input frequencies
C. The original frequencies and the sum and difference frequencies
D. 1.414 and 0.707 times the input frequency

3F64
What are the advantages of the frequency-conversion process?

A. Automatic squelching and increased selectivity
B. Increased selectivity and optimal tuned-circuit design
C. Automatic soft limiting and automatic squelching
D. Automatic detection in the RF amplifier and increased selectivity

3F65
What occurs in a receiver when an excessive amount of signal energy reaches the mixer circuit?

A. Spurious mixer products are generated
B. Mixer blanking occurs
C. Automatic limiting occurs
D. A beat frequency is generated

3F66
How much gain should be used in the RF amplifier stage of a receiver?

A. As much gain as possible short of self oscillation
B. Sufficient gain to allow weak signals to overcome noise generated in the first mixer stage
C. Sufficient gain to keep weak signals below the noise of the first mixer stage
D. It depends on the amplification factor of the first IF stage

3F67
Why should the RF amplifier stage of a receiver only have sufficient gain to allow weak signals to overcome noise generated in the first mixer stage?

A. To prevent the sum and difference frequencies from being generated
B. To prevent bleed-through of the desired signal
C. To prevent the generation of spurious mixer products
D. To prevent bleed-through of the local oscillator

3F68
What is the primary purpose of an RF amplifier in a receiver?

A. To provide most of the receiver gain
B. To vary the receiver image rejection by utilizing the AGC
C. To improve the receiver's noise figure
D. To develop the AGC voltage

3F69
What is an i-f amplifier stage?

A. A fixed-tuned pass-band amplifier
B. A receiver demodulator
C. A receiver filter
D. A buffer oscillator

3F70
What factors should be considered when selecting an intermediate frequency?

A. Cross-modulation distortion and interference
B. Interference to other services
C. Image rejection and selectivity
D. Noise figure and distortion

3F71
What is the primary purpose of the first i-f amplifier stage in a receiver?

A. Noise figure performance
B. Tune out cross-modulation distortion
C. Dynamic response
D. Selectivity

3F72
What is the primary purpose of the final i-f amplifier stage in a receiver?

A. Dynamic response
B. Gain
C. Noise figure performance
D. Bypass undesired signals

3F73
What is a flip-flop circuit?

A. A binary sequential logic element with one stable state
B. A binary sequential logic element with eight stable states
C. A binary sequential logic element with four stable states
D. A binary sequential logic element with two stable states

3F74
How many bits of information can be stored in a single flip-flop circuit?

A. 1
B. 2
C. 3
D. 4

3F75
What is a bistable multivibrator circuit?

A. An "AND" gate
B. An "OR" gate
C. A flip-flop
D. A clock

3F76
How many output changes are obtained for every two trigger pulses applied to the input of a bistable T flip-flop circuit?

A. No output level changes
B. One output level change
C. Two output level changes
D. Four output level changes

3F77
The frequency of an AC signal can be divided electronically by what type of digital circuit?

A. A free-running multivibrator
B. An OR gate
C. A bistable multivibrator
D. An astable multivibrator

3F78
What type of digital IC is also known as a latch?

A. A decade counter
B. An OR gate
C. A flip-flop
D. An op-amp

3F79
How many flip-flops are required to divide a signal frequency by 4?

A. 1
B. 2
C. 4
D. 8

3F80
What is an astable multivibrator?

A. A circuit that alternates between two stable states
B. A circuit that alternates between a stable state and an unstable state
C. A circuit set to block either a 0 pulse or a 1 pulse and pass the other

D. A circuit that alternates between two unstable states

3F81
What is a monostable multivibrator?

A. A circuit that can be switched momentarily to the opposite binary state and then returns after a set time to its original state
B. A "clock" circuit that produces a continuous square wave oscillating between 1 and 0
C. A circuit designed to store one bit of data in either the 0 or the 1 configuration
D. A circuit that maintains a constant output voltage, regardless of variations in the input voltage

3F82
What is an AND gate?

A. A circuit that produces a logic "1" at its output only if all inputs are logic "1"
B. A circuit that produces a logic "0" at its output only if all inputs are logic "1"
C. A circuit that produces a logic "1" at its output if only one input is a logic "1"
D. A circuit that produces a logic "1" at its output if all inputs are logic "0"

3F83
What is a NAND gate?

A. A circuit that produces a logic "0" at its output only when all inputs are logic "0"
B. A circuit that produces a logic "1" at its output only when all inputs are logic "1"
C. A circuit that produces a logic "0" at its output if some but not all of its inputs are logic "1"
D. A circuit that produces a logic "0" at its output only when all inputs are logic "1"

3F84
What is an OR gate?

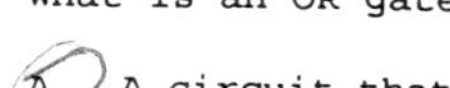

A. A circuit that produces a logic "1" at its output if any input is logic "1"
B. A circuit that produces a logic "0" at its output if any input is logic "1"
C. A circuit that produces a logic "0" at its output if all inputs are logic "1"
D. A circuit that produces a logic "1" at its output if all inputs are logic "0"

3F85
What is a NOR gate?

A. A circuit that produces a logic "0" at its output only if all inputs are logic "0"
B. A circuit that produces a logic "1" at its output only if all inputs are logic "1"
C. A circuit that produces a logic "0" at its output if any or all inputs are logic "1"
D. A circuit that produces a logic "1" at its output if some but not all of its inputs are logic "1"

3F86
What is a NOT gate?

A. A circuit that produces a logic "0" at its output when the input is logic "1" and vice versa
B. A circuit that does not allow data transmission when its input is high
C. A circuit that allows data transmission only when its input is high
D. A circuit that produces a logic "1" at its output when the input is logic "1" and vice versa

3F87
What is a truth table?

A. A table of logic symbols that indicate the high logic states of an op-amp
B. A diagram showing logic states when the digital device's output is true
C. A list of input combinations and their corresponding outputs that characterizes a digital device's function
D. A table of logic symbols that indicates the low logic states of an op-amp

3F88
In a positive-logic circuit, what level is used to represent a logic 1?

A. A low level
B. A positive-transition level
C. A negative-transition level
D. A high level

3F89
In a positive-logic circuit, what level is used to represent a logic 0?

A. A low level
B. A positive-transition level
C. A negative-transition level
D. A high level

3F90
In a negative-logic circuit, what level is used to represent a logic 1?

A. A low level
B. A positive-transition level
C. A negative-transition level
D. A high level

3F91
In a negative-logic circuit, what level is used to represent a logic 0?

A. A low level
B. A positive-transition level
C. A negative-transition level
D. A high level

3F92
What is a crystal-controlled marker generator?

A. A low-stability oscillator that "sweeps" through a band of frequencies
B. An oscillator often used in aircraft to determine the craft's location relative to the inner and outer markers at airports
C. A high-stability oscillator whose output frequency and amplitude can be varied over a wide range
D. A high-stability oscillator that generates a series of reference signals at known frequency intervals

3F93
What additional circuitry is required in a 100-kHz crystal-controlled marker generator to provide markers at 50 and 25 kHz?

A. An emitter-follower
B. Two frequency multipliers
C. Two flip-flops
D. A voltage divider

3F94
What is the purpose of a prescaler circuit?

A. It converts the output of a JK flip-flop to that of an RS flip-flop
B. It multiplies an HF signal so a low-frequency counter can display the operating frequency
C. It prevents oscillation in a low frequency counter circuit
D. It divides an HF signal so a low-frequency counter can display the operating frequency

3F95
What does the accuracy of a frequency counter depend on?

A. The internal crystal reference
B. A voltage-regulated power supply with an unvarying output
C. Accuracy of the AC input frequency to the power supply
D. Proper balancing of the power-supply diodes

3F96
How many states does a decade counter digital IC have?

A. 6
B. 10
C. 15
D. 20

3F97
What is the function of a decade counter digital IC?

A. Decode a decimal number for display on a seven-segment LED display
B. Produce one output pulse for every ten input pulses
C. Produce ten output pulses for every input pulse
D. Add two decimal numbers

3F98
What are the advantages of using an op-amp instead of LC elements in an audio filter?

A. Op-amps are more rugged and can withstand more abuse than can LC elements
B. Op-amps are fixed at one frequency
C. Op-amps are available in more styles and types than are LC elements
D. Op-amps exhibit gain rather than insertion loss

3F99
What determines the gain and frequency characteristics of an op-amp RC active filter?

A. Values of capacitances and resistances built into the op-amp
B. Values of capacitances and resistances external to the op-amp
C. Voltage and frequency of DC input to the op-amp power supply
D. Regulated DC voltage output from the op-amp power supply

3F100
What are the principle uses of an op-amp RC active filter?

A. Op-amp circuits are used as high-pass filters to block RFI at the input to receivers
B. Op-amp circuits are used as low-pass filters between transmitters and transmission lines
C. Op-amp circuits are used as filters for smoothing power-supply output
D. Op-amp circuits are used as audio filters for receivers

3F101
What type of capacitors should be used in an op-amp RC active filter circuit?

A. Electrolytic
B. Disc ceramic
C. Polystyrene
D. Paper dielectric

3F102
How can unwanted ringing and audio instability be prevented in a multisection op-amp RC audio filter circuit?

A. Restrict both gain and Q
B. Restrict gain, but increase Q
C. Restrict Q, but increase gain
D. Increase both gain and Q

3F103
Where should an op-amp RC active audio filter be placed in a receiver?

A. In the IF strip, immediately before the detector
B. In the audio circuitry immediately before the speaker or phone jack
C. Between the balanced modulator and frequency multiplier
D. In the low-level audio stages

3F104
What parameter must be selected when designing an audio filter using an op-amp?

A. Bandpass characteristics
B. Desired current gain
C. Temperature coefficient
D. Output-offset overshoot

3F105
What two factors determine the sensitivity of a receiver?

A. Dynamic range and third-order intercept
B. Cost and availability
C. Intermodulation distortion and dynamic range
D. Bandwidth and noise figure

3F106
What is the limiting condition for sensitivity in a communications receiver?

A. The noise floor of the receiver
B. The power-supply output ripple
C. The two-tone intermodulation distortion
D. The input impedance to the detector

3F107
What is the theoretical minimum noise floor of a receiver with a 400-hertz bandwidth?

A. -141 dBm
B. -148 dBm
C. -174 dBm
D. -180 dBm

3F108
How can selectivity be achieved in the front-end circuitry of a communications receiver?

A. By using an audio filter
B. By using a preselector
C. By using an additional RF amplifier stage
D. By using an additional IF amplifier stage

3F109
A receiver selectivity of 2.4 kHz in the IF circuitry is optimum for what type of signals?

A. CW
B. SSB voice
C. Double-sideband AM voice
D. FSK RTTY

3F110
What occurs during CW reception if too narrow a filter bandwidth is used in the IF stage of a receiver?

A. Undesired signals will reach the audio stage
B. Output-offset overshoot
C. Cross-modulation distortion
D. Filter ringing

3F111
To find the supply power with only a voltmeter, measure between:

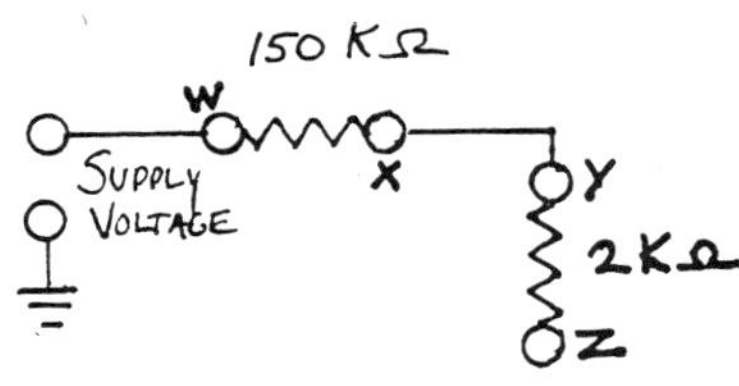

A. Z to W voltage times 150,000 divided by 2000
B. Y to Z voltage squared divided by 2000
C. W to Y voltage divided by 150
D. W to X voltage divided by 150,000 times W to Z voltage

3F112
A receiver selectivity of 10 kHz in the IF circuitry is optimum for what type of signals?

A. SSB voice
B. Double-sideband AM
C. CW
D. FSK RTTY

3F113
What degree of selectivity is desirable in the IF circuitry of a single-sideband phone receiver?

A. 1 kHz
B. 2.4 kHz
C. 4.2 kHz
D. 4.8 kHz

3F114
What is an undesirable effect of using too wide a filter bandwidth in the IF section of a receiver?

A. Output-offset overshoot
B. Undesired signals will reach the audio stage
C. Thermal-noise distortion
D. Filter ringing

3F115
How should the filter bandwidth of a receiver IF section compare with the bandwidth of a received signal?

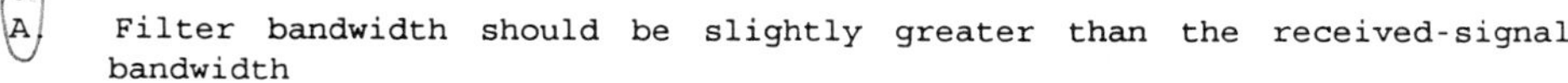
A. Filter bandwidth should be slightly greater than the received-signal bandwidth
B. Filter bandwidth should be approximately half the received-signal bandwidth
C. Filter bandwidth should be approximately two times the received-signal bandwidth
D. Filter bandwidth should be approximately four times the received-signal bandwidth

3F116
What degree of selectivity is desirable in the IF circuitry of an FM-phone receiver?

A. 1 kHz
B. 2.4 kHz
C. 4.2 kHz
D. 15 kHz

3F117
How can selectivity be achieved in the IF circuitry of a communications receiver?

A. Incorporate a means of varying the supply voltage to the local oscillator circuitry
B. Replace the standard JFET mixer with a bipolar transistor followed by a capacitor of the proper value
C. Remove AGC action from the IF stage and confine it to the audio stage only
D. Incorporate a high-Q filter

3F118
What is meant by the dynamic range of a communications receiver?

A. The number of kHz between the lowest and the highest frequency to which the receiver can be tuned
B. The maximum possible undistorted audio output of the receiver, referenced to one milliwatt
C. The ratio between the minimum discernible signal and the largest tolerable signal without causing audible distortion products
D. The difference between the lowest-frequency signal and the highest-frequency signal detectable without moving the tuning knob

3F119
What is the term for the ratio between the largest tolerable receiver input signal and the minimum discernible signal?

A. Intermodulation distortion
B. Noise floor
C. Noise figure
D. Dynamic range

3F120
What type of problems are caused by poor dynamic range in a communications receiver?

A. Cross-modulation of the desired signal and desensitization from strong adjacent signals
B. Oscillator instability requiring frequent retuning, and loss of ability to recover the opposite sideband, should it be transmitted
C. Cross-modulation of the desired signal and insufficient audio power to operate the speaker
D. Oscillator instability and severe audio distortion of all but the strongest received signals

3F121
The ability of a communications receiver to perform well in the presence of strong signals outside the band of interest is indicated by what parameter?

A. Noise figure
B. Blocking dynamic range
C. Signal-to-noise ratio
D. Audio output

3F122
What is meant by the term noise figure of a communications receiver?

A. The level of noise entering the receiver from the antenna
B. The relative strength of a received signal 3 kHz removed from the carrier frequency
C. The level of noise generated in the front end and succeeding stages of a receiver
D. The ability of a receiver to reject unwanted signals at frequencies close to the desired one

3F123
Which stage of a receiver primarily establishes its noise figure?

A. The audio stage
B. The IF strip
C. The RF stage
D. The local oscillator

3F124
What is an inverting op-amp circuit?

A. An operational amplifier circuit connected such that the input and output signals are 180 degrees out of phase
B. An operational amplifier circuit connected such that the input and output signals are in phase
C. An operational amplifier circuit connected such that the input and output signals are 90 degrees out of phase
D. An operational amplifier circuit connected such that the input impedance is held at zero, while the output impedance is high

3F125
What is a noninverting op-amp circuit?

A. An operational amplifier circuit connected such that the input and output signals are 180 degrees out of phase
B. An operational amplifier circuit connected such that the input and output signals are in phase
C. An operational amplifier circuit connected such that the input and output signals are 90 degrees out of phase
D. An operational amplifier circuit connected such that the input impedance is held at zero while the output impedance is high

3F126
How does the gain of a theoretically ideal operational amplifier vary with frequency?

A. The gain increases linearly with increasing frequency
B. The gain decreases linearly with increasing frequency
C. The gain decreases logarithmically with increasing frequency
D. The gain does not vary with frequency

3F127
What determines the input impedance in a FET common-source amplifier?

A. The input impedance is essentially determined by the resistance between the drain and substrate
B. The input impedance is essentially determined by the resistance between the source and drain
C. The input impedance is essentially determined by the gate biasing network
D. The input impedance is essentially determined by the resistance between the source and substrate

3F128
What determines the output impedance in a FET common-source amplifier?

A. The output impedance is essentially determined by the drain resistor
B. The output impedance is essentially determined by the input impedance of the FET
C. The output impedance is essentially determined by the drain supply voltage
D. The output impedance is essentially determined by the gate supply voltage

3F129
What is the purpose of a bypass capacitor?

A. It increases the resonant frequency of the circuit
B. It removes direct current from the circuit by shunting DC to ground
C. It removes alternating current by providing a low impedance path to ground
D. It acts as a voltage divider

3F130
What is the purpose of a coupling capacitor?

A. It blocks direct current and passes alternating current
B. It blocks alternating current and passes direct current
C. It increases the resonant frequency of the circuit
D. It decreases the resonant frequency of the circuit

3F131
What condition must exist for a circuit to oscillate?

A. It must have a gain of less than 1
B. It must be neutralized
C. It must have positive feedback sufficient to overcome losses
D. It must have negative feedback sufficient to cancel the input

3F132
What is the voltage drop across R1?

A. 9 volts
B. 7 volts
C. 5 volts
D. 3 volts

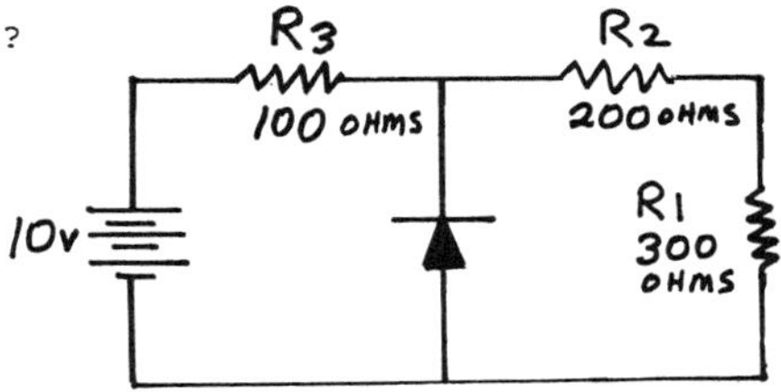

3F133
What is the voltage drop across R1?

A. 1.2 volts
B. 2.4 volts
C. 3.7 volts
D. 9 volts

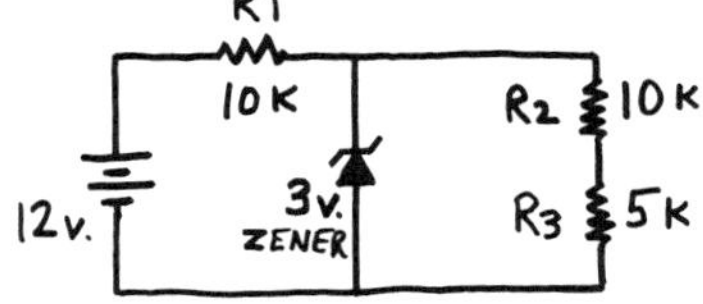

3F134
In a properly operating marine transmitter, if the power supply bleeder resistor opens:

A. Short circuit of supply voltage due to overload
B. Regulation would decrease
C. Next stage would fail due to short circuit
D. Filter capacitors might short from voltage surge

3F135
Which of the following can occur that would least affect this circuit?

A. C1 shorts
B. C1 opens
C. C3 shorts
D. C18 opens

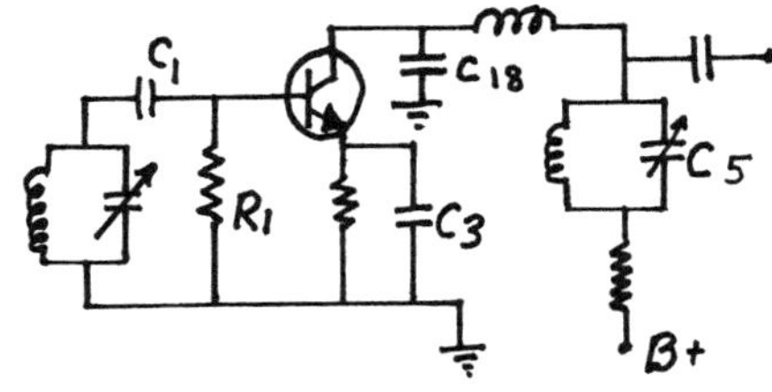

3F136
When S1 is closed, lights L1 and L2 go on. What is the condition of both lamps when both S1 and S2 are closed?

A. both lamps stay on
B. L1 turns off; L2 stays on
C. both lamps turn off
D. L1 stays on; L2 turns off

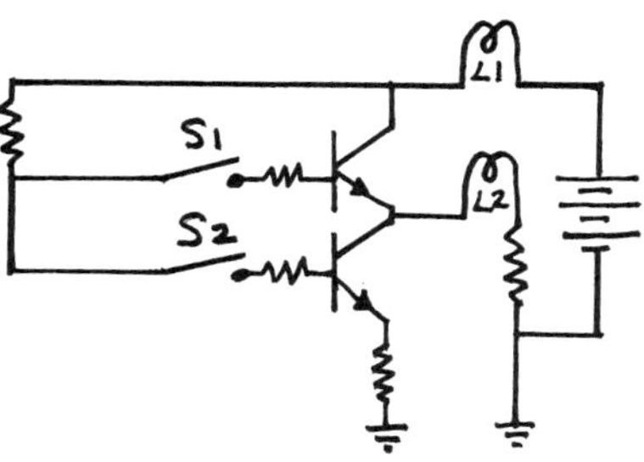

3F137
If S1 is closed both lamps light, what happens when S1 and S2 are closed?

A. L1 and L2 are off
B. L1 is on and L2 is flashing
C. L1 is off and L2 is on
D. L1 is on and L2 is off

(Refer to diagram above)

3F138
How can we correct the defect, if any, in this voltage doubler circuit?

A. Omit C1
B. Reverse polarity signs
C. Ground X
D. Reverse polarity on C1

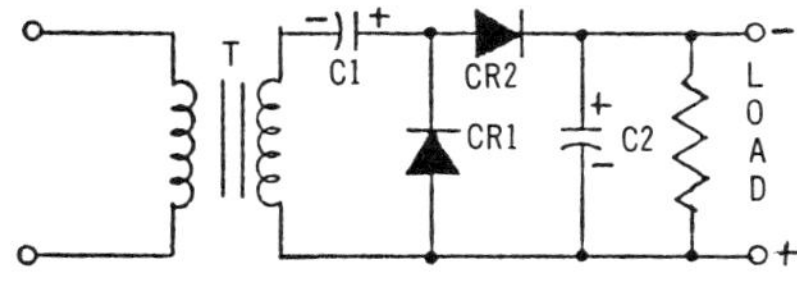

3F139
What change is needed in order to correct the grounded emitter amplifier shown?

A. no change is necessary
B. polarities of emitter-base battery should be reversed
C. polarities of collector-base battery should be reversed
D. point A should be replaced with a low value capacitor

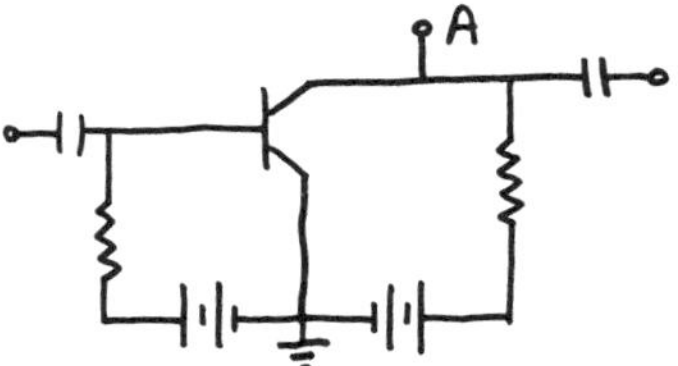

Test 3-G Element Three
Signals and Emissions

3G1
What is emission A3C?

A. Facsimile
B. RTTY
C. ATV
D. Slow Scan TV

3G2
What type of emission is produced when an amplitude modulated transmitter is modulated by a facsimile signal?

A. A3F
B. A3C
C. F3F
D. F3C

3G3
What is facsimile?

A. The transmission of tone-modulated telegraphy
B. The transmission of a pattern of printed characters designed to form a picture
C. The transmission of printed pictures by electrical means
D. The transmission of moving pictures by electrical means

3G4
What is emission F3C?

A. Voice transmission
B. Slow Scan TV
C. RTTY
D. Facsimile

3G5
What type of emission is produced when a frequency modulated transmitter is modulated by a facsimile signal?

A. F3C
B. A3C
C. F3F
D. A3F

3G6
What is emission A3F?

A. RTTY
B. Television
C. SSB
D. Modulated CW

3G7
What type of emission is produced when an amplitude modulated transmitter is modulated by a television signal?

A. F3F
B. A3F
C. A3C
D. F3C

3G8
What is emission F3F?

A. Modulated CW
B. Facsimile
C. RTTY
D. Television

3G9
What type of emission is produced when a frequency modulated transmitter is modulated by a television signal?

A. A3F
B. A3C
C. F3F
D. F3C

3G10
How can an FM-phone signal be produced?

A. By modulating the supply voltage to a class-B amplifier
B. By modulating the supply voltage to a class-C amplifier
C. By using a reactance modulator on an oscillator
D. By using a balanced modulator on an oscillator

3G11
How can a double-sideband phone signal be produced?

A. By using a reactance modulator on an oscillator
B. By varying the voltage to the varactor in an oscillator circuit
C. By using a phase detector, oscillator and filter in a feedback loop
D. By modulating the plate supply voltage to a class C amplifier

3G12
How can a single-sideband phone signal be produced?

A. By producing a double sideband signal with a balanced modulator and then removing the unwanted sideband by filtering
B. By producing a double sideband signal with a balanced modulator and then removing the unwanted sideband by heterodyning
C. By producing a double sideband signal with a balanced modulator and then removing the unwanted sideband by mixing
D. By producing a double sideband signal with a balanced modulator and then removing the unwanted sideband by neutralization

3G13
What is meant by the term deviation ratio?

A. The ratio of the audio modulating frequency to the center carrier frequency
B. The ratio of the maximum carrier frequency deviation to the highest audio modulating frequency
C. The ratio of the carrier center frequency to the audio modulating frequency
D. The ratio of the highest audio modulating frequency to the average audio modulating frequency

3G14
In an FM-phone signal, what is the term for the maximum deviation from the carrier frequency divided by the maximum audio modulating frequency?

A. Deviation index
B. Modulation index
C. Deviation ratio
D. Modulation ratio

3G15
What is the deviation ratio for an FM-phone signal having a maximum frequency swing of plus or minus 5 kHz and accepting a maximum modulation rate of 3 kHz?

A. 60
B. 0.16
C. 0.6
D. 1.66

3G16
What is the deviation ratio of an FM-phone signal having a maximum frequency swing of plus or minus 7.5 kHz and accepting a maximum modulation rate of 3.5 kHz?

A. 2.14
B. 0.214
C. 0.47
D. 47

3G17
What is meant by the term modulation index?

A. The processor index
B. The ratio between the deviation of a frequency modulated signal and the modulating frequency
C. The FM signal-to-noise ratio
D. The ratio of the maximum carrier frequency deviation to the highest audio modulating frequency

3G18
In an FM-phone signal, what is the term for the ratio between the deviation of the frequency-modulated signal and the modulating frequency?

A. FM compressibility
B. Quieting index
C. Percentage of modulation
D. Modulation index

3G19
How does the modulation index of a phase-modulated emission vary with the modulated frequency?

A. The modulation index increases as the RF carrier frequency (the modulated frequency) increases
B. The modulation index decreases as the RF carrier frequency (the modulated frequency) increases
C. The modulation index varies with the square root of the RF carrier frequency (the modulated frequency)
D. The modulation index does not depend on the RF carrier frequency (the modulated frequency)

3G20
In an FM-phone signal having a maximum frequency deviation of 3.0 kHz either side of the carrier frequency, what is the modulation index when the modulating frequency is 1.0 kHz?

A. 3
B. 0.3
C. 3000
D. 1000

3G21
What is the modulation index of an FM-phone transmitter producing an instantaneous carrier deviation of 6 kHz when modulated with a 2 kHz modulating frequency?

A. 6000
B. 3
C. 2000
D. 1/3

3G22
What are electromagnetic waves?

A. Alternating currents in the core of an electromagnet
B. A wave consisting of two electric fields at right angles to each other
C. A wave consisting of an electric field and a magnetic field at right angles to each other
D. A wave consisting of two magnetic fields at right angles to each other

3G23
What is a wave front?

A. A voltage pulse in a conductor
B. A current pulse in a conductor
C. A voltage pulse across a resistor
D. A fixed point in an electromagnetic wave

3G24
At what speed do electromagnetic waves travel in free space?

A. Approximately 300 million meters per second
B. Approximately 468 million meters per second
C. Approximately 186,300 feet per second
D. Approximately 300 million miles per second

3G25
What are the two interrelated fields considered to make up an electromagnetic wave?

A. An electric field and a current field
B. An electric field and a magnetic field
C. An electric field and a voltage field
D. A voltage field and a current field

3G26
Why do electromagnetic waves not penetrate a good conductor to any great extent?

A. The electromagnetic field induces currents in the insulator
B. The oxide on the conductor surface acts as a shield
C. Because of Eddy currents
D. The resistivity of the conductor dissipates the field

3G27
What is meant by referring to electromagnetic waves as horizontally polarized?

A. The electric field is parallel to the earth
B. The magnetic field is parallel to the earth
C. Both the electric and magnetic fields are horizontal
D. Both the electric and magnetic fields are vertical

3G28
What is meant by referring to electromagnetic waves as having circular polarization?

A. The electric field is bent into a circular shape
B. The electric field rotates
C. The electromagnetic wave continues to circle the earth
D. The electromagnetic wave has been generated by a quad antenna

3G29
When the electric field is perpendicular to the surface of the earth, what is the polarization of the electromagnetic wave?

A. Circular
B. Horizontal
C. Vertical
D. Elliptical

3G30
When the magnetic field is parallel to the surface of the earth, what is the polarization of the electromagnetic wave?

A. Circular
B. Horizontal
C. Elliptical
D. Vertical

3G31
When the magnetic field is perpendicular to the surface of the earth, what is the polarization of the electromagnetic field?

A. Horizontal
B. Circular
C. Elliptical
D. Vertical

3G32
When the electric field is parallel to the surface of the earth, what is the polarization of the electromagnetic wave?

A. Vertical
B. Horizontal
C. Circular
D. Elliptical

3G33
What is a sine wave?

A. A constant-voltage, varying-current wave
B. A wave whose amplitude at any given instant can be represented by a point on a wheel rotating at a uniform speed
C. A wave following the laws of the trigonometric tangent function
D. A wave whose polarity changes in a random manner

3G34
If a sine wave begins from above or below the zero axis, how many times will it cross the zero axis in one complete cycle?

A. 180 times
B. 4 times
C. 2 times
D. 360 times

3G35
How many degrees are there in one complete sine wave cycle?

A. 90 degrees
B. 270 degrees
C. 180 degrees
D. 360 degrees

3G36
What is the period of a wave?

A. The time required to complete one cycle
B. The number of degrees in one cycle
C. The number of zero crossings in one cycle
D. The amplitude of the wave

3G37
What is a square wave?

A. A wave with only 300 degrees in one cycle
B. A wave which abruptly changes back and forth between two voltage levels and which remains an equal time at each level
C. A wave that makes four zero crossings per cycle
D. A wave in which the positive and negative excursions occupy unequal portions of the cycle time

3G38
What is a wave called which abruptly changes back and forth between two voltage levels and which remains an equal time at each level?

A. A sine wave
B. A cosine wave
C. A square wave
D. A rectangular wave

3G39
Which sine waves make up a square wave?

A. 0.707 times the fundamental frequency
B. The fundamental frequency and all odd and even harmonics
C. The fundamental frequency and all even harmonics
D. The fundamental frequency and all odd harmonics

3G40
What type of wave is made up of sine waves of the fundamental frequency and all the odd harmonics?

A. Square wave
B. Sine wave
C. Cosine wave
D. Tangent wave

3G41
What is a sawtooth wave?

A. A wave that alternates between two values and spends an equal time at each level
B. A wave with a straight line rise time faster than the fall time (or vice versa)
C. A wave that produces a phase angle tangent to the unit circle
D. A wave whose amplitude at any given instant can be represented by a point on a wheel rotating at a uniform speed

3G42
What type of wave is characterized by a rise time significantly faster than the fall time (or vice versa)?

A. A cosine wave
B. A square wave
C. A sawtooth wave
D. A sine wave

3G43
Which sine waves make up a sawtooth wave?

A. The fundamental frequency and all prime harmonics
B. The fundamental frequency and all even harmonics
C. The fundamental frequency and all odd harmonics
D. The fundamental frequency and all harmonics

3G44
What type of wave is made up of sine waves at the fundamental frequency and all the harmonics?

A. A sawtooth wave
B. A square wave
C. A sine wave
D. A cosine wave

3G45
What is the meaning of the term root mean square value of an AC voltage?

A. The value of an AC voltage found by squaring the average value of the peak AC voltage
B. The value of a DC voltage that would cause the same heating effect in a given resistor as a peak AC voltage
C. The value of an AC voltage that would cause the same heating effect in a given resistor as a DC voltage of the same value
D. The value of an AC voltage found by taking the square root of the average AC value

3G46
What is the term used in reference to a DC voltage that would cause the same heating in a resistor as a certain value of AC voltage?

A. Cosine voltage
B. Power factor
C. Root mean square
D. Average voltage

3G47
What would be the most accurate way of determining the RMS voltage of a complex waveform?

A. By using a grid dip meter
B. By measuring the voltage with a D'Arsonval meter
C. By using an absorption wavemeter
D. By measuring the heating effect in a known resistor

3G48
What is the RMS voltage at a common household electrical power outlet?

A. 117-V AC
B. 331-V AC
C. 82.7-V AC
D. 165.5-V AC

3G49
What is the peak voltage at a common household electrical outlet?

A. 234 volts
B. 165.5 volts
C. 117 volts
D. 331 volts

3G50
What is the peak-to-peak voltage at a common household electrical outlet?

A. 234 volts
B. 117 volts
C. 331 volts
D. 165.5 volts

3G51
What is RMS voltage of a 165-volt peak pure sine wave?

A. 233-V AC
B. 330-V AC
C. 58.3-V AC
D. 117-V AC

3G52
What is the RMS value of a 331-volt peak-to-peak pure sine wave?

A. 117-V AC
B. 165-V AC
C. 234-V AC
D. 300-V AC

3G53
For many types of voices, what is the ratio of PEP to average power during a modulation peak in a single-sideband phone signal?

A. Approximately 1.0 to 1
B. Approximately 25 to 1
C. Approximately 2.5 to 1
D. Approximately 100 to 1

3G54
In a single-sideband phone signal, what determines the PEP-to-average power ratio?

A. The frequency of the modulating signal
B. The degree of carrier suppression
C. The speech characteristics
D. The amplifier power

3G55
What is the approximate DC input power to a Class B RF power amplifier stage in an FM-phone transmitter when the PEP output power is 1500 watts?

A. Approximately 900 watts
B. Approximately 1765 watts
C. Approximately 2500 watts
D. Approximately 3000 watts

3G56
What is the approximate DC input power to a Class C RF power amplifier stage in a RTTY transmitter when the PEP output power is 1000 watts?

A. Approximately 850 watts
B. Approximately 1250 watts
C. Approximately 1667 watts
D. Approximately 2000 watts

3G57
What is the approximate DC input power to a Class AB RF power amplifier stage in an unmodulated carrier transmitter when the PEP output power is 500 watts?

A. Approximately 250 watts
B. Approximately 600 watts
C. Approximately 800 watts
D. Approximately 1000 watts

3G58
Where is the noise generated which primarily determines the signal-to-noise ratio in a VHF (150 MHz) marine band receiver?

A. In the receiver front end
B. Man-made noise
C. In the atmosphere
D. In the ionosphere

3G59
In a pulse-width modulation system, what parameter does the modulating signal vary?

A. Pulse duration
B. Pulse frequency
C. Pulse amplitude
D. Pulse intensity

3G60
What is the type of modulation in which the modulating signal varies the duration of the transmitted pulse?

A. Amplitude modulation
B. Frequency modulation
C. Pulse-width modulation
D. Pulse-height modulation

3G61
In a pulse-position modulation system, what parameter does the modulating signal vary?

A. The number of pulses per second
B. Both the frequency and amplitude of the pulses
C. The duration of the pulses
D. The time at which each pulse occurs

3G62
Why is the transmitter peak power in a pulse modulation system much greater than its average power?

A. The signal duty cycle is less than 100%
B. The signal reaches peak amplitude only when voice modulated
C. The signal reaches peak amplitude only when voltage spikes are generated within the modulator
D. The signal reaches peak amplitude only when the pulses are also amplitude modulated

3G63
What is one way that voice is transmitted in a pulse-width modulation system?

A. A standard pulse is varied in amplitude by an amount depending on the voice waveform at that instant
B. The position of a standard pulse is varied by an amount depending on the voice waveform at that instant
C. A standard pulse is varied in duration by an amount depending on the voice waveform at that instant
D. The number of standard pulses per second varies depending on the voice waveform at that instant

3G64
The International Organization for Standardization has developed a seven-level reference model for a packet-radio communications structure. What level is responsible for the actual transmission of data and handshaking signals?

A. The physical layer
B. The transport layer
C. The communications layer
D. The synchronization layer

3G65
The International Organization for Standardization has developed a seven-level reference model for a packet-radio communications structure. What level arranges the bits into frames and controls data flow?

A. The transport layer
B. The link layer
C. The communications layer
D. The synchronization layer

3G66
What is one advantage of using the ASCII code, with its larger character set, instead of the Baudot code?

A. ASCII includes built-in error-correction features
B. ASCII characters contain fewer information bits than Baudot characters
C. It is possible to transmit upper and lower case text
D. The larger character set allows store-and-forward control characters to be added to a message

3G67
What is the duration of a 45-baud Baudot RTTY data pulse?

A. 11 milliseconds
B. 40 milliseconds
C. 31 milliseconds
D. 22 milliseconds

3G68
What is the duration of a 45-baud Baudot RTTY start pulse?

A. 11 milliseconds
B. 22 milliseconds
C. 31 milliseconds
D. 40 milliseconds

3G69
What is the duration of a 45-baud Baudot RTTY stop pulse?

A. 11 milliseconds
B. 18 milliseconds
C. 31 milliseconds
D. 40 milliseconds

3G70
What is the necessary bandwidth of a 170-hertz shift, 45-baud Baudot emission F1B transmission?

A. 45 Hz
B. 249 Hz
C. 442 Hz
D. 600 Hz

3G71
What is the necessary bandwidth of a 170-hertz shift, 45-baud Baudot emission J2B transmission?

A. 45 Hz
B. 249 Hz
C. 442 Hz
D. 600 Hz

3G72
What is the necessary bandwidth of a 170-hertz shift, 74-baud Baudot emission F1B transmission?

A. 250 Hz
B. 278 Hz
C. 442 Hz
D. 600 Hz

3G73
What is the necessary bandwidth of a 170-hertz shift, 74-baud Baudot emission J2B transmission?

A. 250 Hz
B. 278 Hz
C. 442 Hz
D. 600 Hz

3G74
What is the necessary bandwidth of a 1000-hertz shift, 1200-baud ASCII emission F1D transmission?

A. 1000 Hz
B. 1200 Hz
C. 440 Hz
D. 2400 Hz

3G75
What is the necessary bandwidth of a 4800-hertz frequency shift, 9600-baud ASCII emission F1D transmission?

A. 15.36 kHz
B. 9.6 kHz
C. 4.8 kHz
D. 5.76 kHz

3G76
What is the necessary bandwidth of a 4800-hertz frequency shift, 9600-baud ASCII emission J2D transmission?

A. 15.36 kHz
B. 9.6 kHz
C. 4.8 kHz
D. 5.76 kHz

3G77
What is the necessary bandwidth of a 170-hertz shift, 110-baud ASCII emission F1B transmission?

A. 304 Hz
B. 314 Hz
C. 608 Hz
D. 628 Hz

3G78
What is the necessary bandwidth of a 170-hertz shift, 110-baud ASCII emission J2B transmission?

A. 304 Hz
B. 314 Hz
C. 608 Hz
D. 628 Hz

3G79
What is the necessary bandwidth of a 170-hertz shift, 300-baud ASCII emission F1D transmission?

A. 0 Hz
B. 0.3 kHz
C. 0.5 kHz
D. 1.0 kHz

3G80
What is the necessary bandwidth for a 170-hertz shift, 300-baud ASCII emission J2D transmission?

A. 0 Hz
B. 0.3 kHz
C. 0.5 kHz
D. 1.0 kHz

3G81
What is amplitude compandored single sideband?

A. Reception of single sideband with a conventional CW receiver
B. Reception of single sideband with a conventional FM receiver
C. Single sideband incorporating speech compression at the transmitter and speech expansion at the receiver
D. Single sideband incorporating speech expansion at the transmitter and speech compression at the receiver

3G82
What is meant by compandoring?

A. Compressing speech at the transmitter and expanding it at the receiver
B. Using an audio-frequency signal to produce pulse-length modulation
C. Combining amplitude and frequency modulation to produce a single-sideband signal
D. Detecting and demodulating a single-sideband signal by converting it to a pulse-modulated signal

3G83
What is the purpose of a pilot tone in an amplitude compandored single sideband system?

A. It permits rapid tuning of a mobile receiver
B. It replaces the suppressed carrier at the receiver
C. It permits rapid change of frequency to escape high-powered interference
D. It acts as a beacon to indicate the present propagation characteristic of the band

3G84
What is the approximate frequency of the pilot tone in an amplitude compandored single sideband system?

A. 1 kHz
B. 5 MHz
C. 455 kHz
D. 3 kHz

3G85
How many more voice transmissions can be packed into a given frequency band for amplitude-compandored single-sideband systems over conventional FM-phone systems?

A. 2
B. 4
C. 8
D. 16

3G86
What term describes a wide-bandwidth communications system in which the RF carrier varies according to some predetermined sequence?

A. Amplitude compandored single sideband
B. SITOR
C. Time-domain frequency modulation
D. Spread spectrum communication

3G87
What is the term used to describe a spread spectrum communications system where the center frequency of a conventional carrier is altered many times per second in accordance with a pseudo-random list of channels?

A. Frequency hopping
B. Direct sequence
C. Time-domain frequency modulation
D. Frequency compandored spread spectrum

3G88
What term is used to describe a spread spectrum communications system in which a very fast binary bit stream is used to shift the phase of an RF carrier?

A. Frequency hopping
B. Direct sequence
C. Binary phase-shift keying
D. Phase compandored spread spectrum

3G89
What is the term for the amplitude of the maximum positive excursion of a signal as viewed on an oscilloscope?

A. Peak-to-peak voltage
B. Inverse peak negative voltage
C. RMS voltage
D. Peak positive voltage

3G90
What is the term for the amplitude of the maximum negative excursion of a signal as viewed on an oscilloscope?

A. Peak-to-peak voltage
B. Inverse peak positive voltage
C. RMS voltage
D. Peak negative voltage

3G91
What is the easiest voltage amplitude dimension to measure by viewing a pure sine wave signal on an oscilloscope?

A. Peak-to-peak voltage
B. RMS voltage
C. Average voltage
D. DC voltage

3G92
What is the relationship between the peak-to-peak voltage and the peak voltage amplitude in a symmetrical wave form?

A. 1:1
B. 2:1
C. 3:1
D. 4:1

3G93
What input-amplitude parameter is valuable in evaluating the signal-handling capability of a Class A amplifier?

A. Peak voltage
B. Average voltage
C. RMS voltage
D. Resting voltage

3G94
If 480 Khz. is radiated from a 1/4 wavelength antenna, what is the 7th harmonic?

A. 3.360 MHz
B. 840 KHz
C. 3350 kHz
D. 480 kHz

3G95

What is the seventh harmonic of a 100 MHz. quarter wavelength antenna?

A. 14.28 MHz.
B. 107 MHz.
C. 149 MHz.
D. 700 MHz.

3G96
What is the seventh harmonic of 2182 kHz. when the transmitter is connected to a half-wave Hertz antenna?

A. 2182 kHz.
B. 15.27 MHz.
C. 311.7 kHz.
D. 7.64 MHz.

3G97
What is the fifth harmonic frequency of a transmitter operating on 480 kHz. with a 1/4 wavelength antenna?

A. 2.4 MHz
B. 240 MHz
C. 600 kHz
D. 1.2 MHz

Test 3-H Element Three
Antennas and Feed Lines

3H1
What is meant by the term antenna gain?

A. The numerical ratio relating the radiated signal strength of an antenna to that of another antenna
B. The ratio of the signal in the forward direction to the signal in the back direction
C. The ratio of the amount of power produced by the antenna compared to the output power of the transmitter
D. The final amplifier gain minus the transmission line losses (including any phasing lines present)

3H2
What is the term for a numerical ratio which relates the performance of one antenna to that of another real or theoretical antenna?

A. Effective radiated power
B. Antenna gain
C. Conversion gain
D. Peak effective power

3H3
What is meant by the term antenna bandwidth?

A. Antenna length divided by the number of elements
B. The frequency range over which an antenna can be expected to perform well
C. The angle between the half-power radiation points
D. The angle formed between two imaginary lines drawn through the ends of the elements

3H4
What is the wavelength of a shorted stub used to absorb even harmonics?

A. 1/2 wavelength
B. 1/3 wavelength
C. 1/4 wavelength
D. 1/8 wavelength

3H5
What is a trap antenna?

A. An antenna for rejecting interfering signals
B. A highly sensitive antenna with maximum gain in all directions
C. An antenna capable of being used on more than one band because of the presence of parallel LC networks
D. An antenna with a large capture area

3H6
What is an advantage of using a trap antenna?

A. It has high directivity in the high-frequency bands
B. It has high gain
C. It minimizes harmonic radiation
D. It may be used for multiband operation

3H7
What is a disadvantage of using a trap antenna?

A. It will radiate harmonics
B. It can only be used for single band operation
C. It is too sharply directional at lower frequencies
D. It must be neutralized

3H8
What is the principle of a trap antenna?

A. Beamwidth may be controlled by non-linear impedances
B. The traps form a high impedance to isolate parts of the antenna
C. The effective radiated power can be increased if the space around the antenna "sees" a high impedance
D. The traps increase the antenna gain

3H9
What is a parasitic element of an antenna?

A. An element polarized 90 degrees opposite the driven element
B. An element dependent on the antenna structure for support
C. An element that receives its excitation from mutual coupling rather than from a transmission line
D. A transmission line that radiates radio-frequency energy

3H10
How does a parasitic element generate an electromagnetic field?

A. By the RF current received from a connected transmission line
B. By interacting with the earth's magnetic field
C. By altering the phase of the current on the driven element
D. By currents induced into the element from a surrounding electric field

3H11
How does the length of the reflector element of a parasitic element beam antenna compare with that of the driven element?

A. It is about 5% longer
B. It is about 5% shorter
C. It is twice as long
D. It is one-half as long

3H12
How does the length of the director element of a parasitic element beam antenna compare with that of the driven element?

A. It is about 5% longer
B. It is about 5% shorter
C. It is one-half as long
D. It is twice as long

3H13
What is meant by the term radiation resistance for an antenna?

A. Losses in the antenna elements and feed line
B. The specific impedance of the antenna
C. An equivalent resistance that would dissipate the same amount of power as that radiated from an antenna
D. The resistance in the trap coils to received signals

3H14
What are the factors that determine the radiation resistance of an antenna?

A. Transmission line length and height of antenna
B. The location of the antenna with respect to nearby objects and the length/diameter ratio of the conductors
C. It is a constant for all antennas since it is a physical constant
D. Sunspot activity and the time of day

3H15
What is a driven element of an antenna?

A. Always the rearmost element
B. Always the forwardmost element
C. The element fed by the transmission line
D. The element connected to the rotator

3H16
What is the usual electrical length of a driven element in a HF beam antenna?

A. 1/4 wavelength
B. 1/2 wavelength
C. 3/4 wavelength
D. 1 wavelength

3H17
What is the term for an antenna element which is supplied power from a transmitter through a transmission line?

A. Driven element
B. Director element
C. Reflector element
D. Parasitic element

3H18
How is antenna "efficiency" computed?

A. Efficiency = (radiation resistance/transmission resistance) X 100%
B. Efficiency = (radiation resistance/total resistance) X 100%
C. Efficiency = (total resistance/radiation resistance) X 100%
D. Efficiency = (effective radiated power/transmitter output) X 100%

3H19
What is the term for the ratio of the radiation resistance of an antenna to the total resistance of the system?

A. Effective radiated power
B. Radiation conversion loss
C. Antenna efficiency
D. Beamwidth

3H20
What is included in the total resistance of an antenna system?

A. Radiation resistance plus space impedance
B. Radiation resistance plus transmission resistance
C. Transmission line resistance plus radiation resistance
D. Radiation resistance plus ohmic resistance

3H21
How can the antenna efficiency of a HF grounded vertical antenna be made comparable to that of a half-wave antenna?

A. By installing a good ground radial system
B. By isolating the coax shield from ground
C. By shortening the vertical
D. By lengthening the vertical

3H22
Why does a half-wave antenna operate at very high efficiency?

A. Because it is non-resonant
B. Because the conductor resistance is low compared to the radiation resistance
C. Because earth-induced currents add to its radiated power
D. Because it has less corona from the element ends than other types of antennas

3H23
What is a folded dipole antenna?

A. A dipole that is one-quarter wavelength long
B. A ground plane antenna
C. A dipole whose ends are connected by another one-half wavelength piece of wire
D. A fictional antenna used in theoretical discussions to replace the radiation resistance

3H24
How does the bandwidth of a folded dipole antenna compare with that of a simple dipole antenna?

A. It is 0.707 times the simple dipole bandwidth
B. It is essentially the same
C. It is less than 50% that of a simple dipole
D. It is greater

3H25
What is the input terminal impedance at the center of a folded dipole antenna?

A. 300 ohms
B. 72 ohms
C. 50 ohms
D. 450 ohms

3H26
What is the meaning of the term velocity factor of a transmission line?

A. The ratio of the characteristic impedance of the line to the terminating impedance
B. The index of shielding for coaxial cable
C. The velocity of the wave on the transmission line multiplied by the velocity of light in a vacuum
D. The velocity of the wave on the transmission line divided by the velocity of light in a vacuum

3H27
What is the term for the ratio of actual velocity at which a signal travels through a line to the speed of light in a vacuum?

A. Velocity factor
B. Characteristic impedance
C. Surge impedance
D. Standing wave ratio

3H28
What is the velocity factor for non-foam dielectric 50 or 75 ohm flexible coaxial cable such as RG 8, 11, 58 and 59?

A. 2.70
B. 0.66
C. 0.30
D. 0.10

3H29
What determines the velocity factor in a transmission line?

A. The termination impedance
B. The line length
C. Dielectrics in the line
D. The center conductor resistivity

3H30
Why is the physical length of a coaxial cable transmission line shorter than its electrical length?

A. Skin effect is less pronounced in the coaxial cable
B. RF energy moves slower along the coaxial cable
C. The surge impedance is higher in the parallel feed line
D. The characteristic impedance is higher in the parallel feed line

3H31
What would be the physical length of a typical coaxial transmission line which is electrically one-quarter wavelength long at 14.1 MHz?

A. 20 meters
B. 3.51 meters
C. 2.33 meters
D. 0.25 meters

3H32
What would be the physical length of a typical coaxial transmission line which is electrically one-quarter wavelength long at 7.2 MHz?

A. 10.5 meters
B. 6.88 meters
C. 24 meters
D. 50 meters

3H33
What is the physical length of a parallel antenna feedline which is electrically one-half wavelength long at 14.10 MHz? (assume a velocity factor of 0.82.)

A. 15 meters
B. 24.3 meters
C. 8.7 meters
D. 70.8 meters

3H34
What is the physical length of a twin lead transmission feedline at 36.5 MHz? (assume a velocity factor of 0.80.)

A. Electrical length times 0.8
B. Electrical length divided by 0.8
C. 80 meters
D. 160 meters

3H35
In a half-wave antenna, where are the current nodes?

A. At the ends
B. At the center
C. Three-quarters of the way from the feed point toward the end
D. One-half of the way from the feed point toward the end

3H36
In a half-wave antenna, where are the voltage nodes?

A. At the ends
B. At the feed point
C. Three-quarters of the way from the feed point toward the end
D. One-half of the way from the feed point toward the end

3H37
At the ends of a half-wave antenna, what values of current and voltage exist compared to the remainder of the antenna?

A. Equal voltage and current
B. Minimum voltage and maximum current
C. Maximum voltage and minimum current
D. Minimum voltage and minimum current

3H38
At the center of a half-wave antenna, what values of voltage and current exist compared to the remainder of the antenna?

A. Equal voltage and current
B. Maximum voltage and minimum current
C. Minimum voltage and minimum current
D. Minimum voltage and maximum current

3H39
What happens to the base feed point of a fixed length mobile antenna as the frequency of operation is lowered?

A. The resistance decreases and the capacitive reactance decreases
B. The resistance decreases and the capacitive reactance increases
C. The resistance increases and the capacitive reactance decreases
D. The resistance increases and the capacitive reactance increases

3H40
Why should an HF mobile antenna loading coil have a high ratio of reactance to resistance?

A. To swamp out harmonics
B. To maximize losses
C. To minimize losses
D. To minimize the Q

3H41
Why is a loading coil often used with an HF mobile antenna?

A. To improve reception
B. To lower the losses
C. To lower the Q
D. To tune out the capacitive reactance

3H42
For a shortened vertical antenna, where should a loading coil be placed to minimize losses and produce the most effective performance?

A. Near the center of the vertical radiator
B. As low as possible on the vertical radiator
C. As close to the transmitter as possible
D. At a voltage node

3H43
What happens to the bandwidth of an antenna as it is shortened through the use of loading coils?

A. It is increased
B. It is decreased
C. No change occurs
D. It becomes flat

3H44
Why are self-resonant antennas popular in many applications?

A. They are very broad banded
B. They have high gain in all azimuthal directions
C. They are the most efficient radiators
D. They require no calculations

3H45
What is an advantage of using top loading in a shortened HF vertical antenna?

A. Lower Q
B. Greater structural strength
C. Higher losses
D. Improved radiation efficiency

3H46
What is an isotropic radiator?

A. A hypothetical, omnidirectional antenna
B. In the northern hemisphere, an antenna whose directive pattern is constant in southern directions
C. An antenna high enough in the air that its directive pattern is substantially unaffected by the ground beneath it
D. An antenna whose directive pattern is substantially unaffected by the spacing of the elements

3H47
When is it useful to refer to an isotropic radiator?

A. When comparing the gains of directional antennas
B. When testing a transmission line for standing wave ratio
C. When (in the northern hemisphere) directing the transmission in a southerly direction
D. When using a dummy load to tune a transmitter

3H48
What theoretical reference antenna provides a comparison for antenna measurements?

A. Quarter-wave vertical
B. Yagi-Uda array
C. Bobtail curtain
D. Isotropic radiator

3H49
What purpose does an isotropic radiator serve?

A. It is used to compare signal strengths (at a distant point) of different transmitters
B. It is used as a reference for antenna gain measurements
C. It is used as a dummy load for tuning transmitters
D. It is used to measure the standing-wave-ratio on a transmission line

3H50
How much gain does a 1/2-wavelength dipole have over an isotropic radiator?

A. About 1.5 dB
B. About 2.1 dB
C. About 3.0 dB
D. About 6.0 dB

3H51
How much gain does an antenna have over a 1/2-wavelength dipole when it has 6 dB gain over an isotropic radiator?

A. About 3.9 dB
B. About 6.0 dB
C. About 8.1 dB
D. About 10.0 dB

3H52
How much gain does an antenna have over a 1/2-wavelength dipole when it has 12 dB gain over an isotropic radiator?

A. About 6.1 dB
B. About 9.9 dB
C. About 12.0 dB
D. About 14.1 dB

3H53
What is the antenna pattern for an isotropic radiator?

A. A figure-8
B. A unidirectional cardioid
C. A parabola
D. A sphere

3H54
What type of directivity pattern does an isotropic radiator have?

A. A figure-8
B. A unidirectional cardioid
C. A parabola
D. A sphere

3H55
What factors determine the receiving antenna gain required at a station in earth operation?

A. Height, transmitter power and antennas of satellite
B. Length of transmission line and impedance match between receiver and transmission line
C. Preamplifier location on transmission line and presence or absence of RF amplifier stages
D. Height of earth antenna and satellite orbit

3H56
What factors determine the EIRP required by a station in earth operation?

A. Satellite antennas and height, satellite receiver sensitivity
B. Path loss, earth antenna gain, signal-to-noise ratio
C. Satellite transmitter power and orientation of ground receiving antenna
D. Elevation of satellite above horizon, signal-to-noise ratio, satellite transmitter power

3H57
What factors determine the EIRP required by a station in telecommand operation?

A. Path loss, earth antenna gain, signal-to-noise ratio
B. Satellite antennas and height, satellite receiver sensitivity
C. Satellite transmitter power and orientation of ground receiving antenna
D. Elevation of satellite above horizon, signal-to-noise ratio, satellite transmitter power

3H58
How does the gain of a parabolic dish type antenna change when the operating frequency is doubled?

A. Gain does not change
B. Gain is multiplied by 0.707
C. Gain increases 6 dB
D. Gain increases 3 dB

3H59
What happens to the beamwidth of an antenna as the gain is increased?

A. The beamwidth increases geometrically as the gain is increased
B. The beamwidth increases arithmetically as the gain is increased
C. The beamwidth is essentially unaffected by the gain of the antenna
D. The beamwidth decreases as the gain is increased

3H60
What is the beamwidth of a symmetrical pattern antenna with a gain of 20 dB as compared to an isotropic radiator?

A. 10.1 degrees
B. 20.3 degrees
C. 45.0 degrees
D. 60.9 degrees

3H61
What is the beamwidth of a symmetrical pattern antenna with a gain of 30 dB as compared to an isotropic radiator?

A. 3.2 degrees
B. 6.4 degrees
C. 37 degrees
D. 60.4 degrees

3H62
What is the beamwidth of a symmetrical pattern antenna with a gain of 15 dB as compared to an isotropic radiator?

A. 72 degrees
B. 52 degrees
C. 36.1 degrees
D. 3.61 degrees

3H63
What is the beamwidth of a symmetrical pattern antenna with a gain of 12 dB as compared to an isotropic radiator?

A. 34.8 degrees
B. 45.0 degrees
C. 58.0 degrees
D. 51.0 degrees

3H64
How is circular polarization produced using linearly-polarized antennas?

A. Stack two yagis, fed 90 degrees out of phase, to form an array with the respective elements in parallel planes
B. Stack two yagis, fed in phase, to form an array with the respective elements in parallel planes
C. Arrange two yagis perpendicular to each other, with the driven elements in the same plane, fed 90 degrees out of phase
D. Arrange two yagis perpendicular to each other, with the driven elements in the same plane, fed in phase

3H65
Why does an antenna system for earth operation (for communications through a satellite) need to have rotators for both azimuth and elevation control?

A. In order to point the antenna above the horizon to avoid terrestrial interference
B. Satellite antennas require two rotators because they are so large and heavy
C. In order to track the satellite as it orbits the earth
D. The elevation rotator points the antenna at the satellite and the azimuth rotator changes the antenna polarization

3H66
What term describes a method used to match a high-impedance transmission line to a lower impedance antenna by connecting the line to the driven element in two places, spaced a fraction of a wavelength on each side of the driven element center?

A. The gamma matching system
B. The delta matching system
C. The omega matching system
D. The stub matching system

3H67
What term describes an unbalanced feed system in which the driven element is fed both at the center of that element and a fraction of a wavelength to one side of center?

A. The gamma matching system
B. The delta matching system
C. The omega matching system
D. The stub matching system

3H68
What term describes a method of antenna impedance matching that uses a short section of transmission line connected to the antenna feed line near the antenna and perpendicular to the feed line?

A. The gamma matching system
B. The delta matching system
C. The omega matching system
D. The stub matching system

3H69
What kind of impedance does a 1/8-wavelength transmission line present to a generator when the line is shorted at the far end?

A. A capacitive reactance
B. The same as the characteristic impedance of the line
C. An inductive reactance
D. The same as the input impedance to the final generator stage

3H70
What kind of impedance does a 1/8-wavelength transmission line present to a generator when the line is open at the far end?

A. The same as the characteristic impedance of the line
B. An inductive reactance
C. A capacitive reactance
D. The same as the input impedance of the final generator stage

3H71
What kind of impedance does a 1/4-wavelength transmission line present to a generator when the line is shorted at the far end?

A. A very high impedance
B. A very low impedance
C. The same as the characteristic impedance of the transmission line
D. The same as the generator output impedance

3H72
What kind of impedance does a 1/4-wavelength transmission line present to a generator when the line is open at the far end?

A. A very high impedance
B. A very low impedance
C. The same as the characteristic impedance of the line
D. The same as the input impedance to the final generator stage

3H73
What kind of impedance does a 3/8-wavelength transmission line present to a generator when the line is shorted at the far end?

A. The same as the characteristic impedance of the line
B. An inductive reactance
C. A capacitive reactance
D. The same as the input impedance to the final generator stage

3H74
What kind of impedance does a 3/8-wavelength transmission line present to a generator when the line is open at the far end?

A. A capacitive reactance
B. The same as the characteristic impedance of the line
C. An inductive reactance
D. The same as the input impedance to the final generator stage

3H75
What kind of impedance does a 1/2-wavelength transmission line present to a generator when the line is shorted at the far end?

A. A very high impedance
B. A very low impedance
C. The same as the characteristic impedance of the line
D. The same as the output impedance of the generator

3H76
What kind of impedance does a 1/2-wavelength transmission line present to a generator when the line is open at the far end?

A. A very high impedance
B. A very low impedance
C. The same as the characteristic impedance of the line
D. The same as the output impedance of the generator

3H77
What is the term used for an equivalent resistance which would dissipate the same amount of energy as that radiated from an antenna?

A. Space resistance
B. Loss resistance
C. Transmission line loss
D. Radiation resistance

3H78
Why is the value of the radiation resistance of an antenna important?

A. Knowing the radiation resistance makes it possible to match impedances for maximum power transfer
B. Knowing the radiation resistance makes it possible to measure the near-field radiation density from a transmitting antenna
C. The value of the radiation resistance represents the front-to-side ratio of the antenna
D. The value of the radiation resistance represents the front-to-back ratio of the antenna

3H79
Adding parasitic elements to an antenna will:

A. Decrease its directional characteristics
B. Decrease its sensitivity
C. Increase its directional characteristics
D. Increase its sensitivity

3H80
What ferrite rod device prevents the formation of reflected waves on a waveguide transmission line?

A. Reflector
B. Isolator
C. Wave-trap
D. SWR refractor

3H81
Frequencies most affected by knife-edge refraction are:

A. Low and medium frequencies
B. High frequencies
C. Very high and ultra high frequencies
D. 100 kHz. to 3.0 MHz.

3H82
When measuring I and V along a 1/2 wave hertz antenna where would you find the points where I and V are maximum and minimum?

A. V and I are high at the ends
B. V and I are high in the middle
C. V and I are uniform throughout the antenna
D. V is maximum at both ends, I is maximum in the middle

3H83
To increase the resonant frequency of a 1/4 wavelength antenna:

A. Add a capacitor
B. Lower capacitor value
C. Cut antenna
D. Add an inductor

3H84
Why are concentric transmission lines sometimes filled with nitrogen?

A. Reduces resistance at high frequencies
B. Prevent water damage underground
C. Keep moisture out and prevent oxidation
D. Reduce microwave line losses

3H85
A vertical 1/4 wave antenna receives signals:

A. In the microwave band
B. In one vertical direction
C. In one horizontal direction
D. Equally from all horizontal directions

3H86
Which of the following represents the best standing wave ratio (SWR)?

A. 1:1
B. 1:1.5
C. 1:3
D. 1:4

3H87
What is the purpose of stacking elements on an antenna?

A. Sharper directional pattern
B. Increased gain
C. Improved bandpass
D. All of the above

3H88
In a quarter-wave Marconi antenna, where is the voltage and current concentrated?

A. At the ends
B. Voltage at the ends, current in the middle
C. Current at the ends, voltage in the middle
D. Evenly throughout

3H89
On a half-wave Hertz antenna:

A. Voltage is maximum at both ends and current is maximum at the center of the antenna
B. Current is maximum at both ends and voltage in the center
C. Voltage and current are uniform throughout the antenna
D. Voltage and current are high at the ends

3H90
What type of antenna is designed for minimum radiation?

A. Dummy antenna
B. Quarter-wave antenna
C. Half-wave antenna
D. Directional antenna

3H91
What is the outcome when you stack antennas at various angles

A. A more omni-directional reception
B. A more uni-directional reception
C. An overall reception signal increase
D. Both a and c

3H92
Adding parasitic elements to a quarter-wavelength antenna will:

A. Reduce its directional characteristics
B. Increase its directional characteristics
C. Increase its sensitivity
D. Reduce its effectiveness

3H93
Ignoring line losses, voltage at a point on a transmission line without standing waves is:

A. Equal to the product of the line current and impedance
B. Equal to the product of the line current and power factor
C. Equal to the product of the line current and the surge impedance
D. Zero at both ends

3H94
Stacking antenna elements:

A. Will suppress odd harmonics
B. Decrease signal to noise ratio
C. Increases sensitivity to weak signals
D. Increases selectivity

3H95
What allows microwaves to pass in only one direction?

A. RF emitter
B. Ferrite isolator
C. Capacitor
D. Varactor-triac

3H96
What would be added to make a receiving antenna more directional?

A. Inductor
B. Capacitor
C. Parasitic elements
D. Height

3H97
Nitrogen is placed in transmission lines to:

A. Improve the 'skin-effect' of microwaves
B. Reduce arcing in the line
C. Reduce the standing wave ratio of the line
D. Prevent moisture from entering the line

3H98
Neglecting line losses, the voltage at any point along a transmission line, having no standing waves, will be equal to the:

A. Transmitter output
B. Product of the line voltage and the surge impedance of the line
C. Product of the line current and the surge impedance of the line
D. Product of the resistance and surge impedance of the line

3H99
Adding a capacitor in series with a Marconi antenna:

A. Increases the antenna circuit resonant frequency
B. Decreases the antenna circuit resonant frequency
C. Blocks the transmission of signals from the antenna
D. Increases the power handling capacity of the antenna

3H100
An antenna is carrying an unmodulated signal, when 100% modulation is impressed, the antenna current:

A. Goes up 50%
B. Goes down one half
C. Stays the same
D. Goes up 22.5%

3H101
An excited 1/2 wavelength antenna produces:

A. Residual fields
B. An electro-magnetic field only
C. Both electro-magnetic and electro-static fields
D. An electro-flux field sometimes

3H102
An antenna which intercepts signals equally from all horizontal directions is:

A. Parabolic
B. Vertical loop
C. Horizontal marconi
D. Vertical 1/4 wave

3H103
Loop-antenna:

A. Is bi-directional
B. Is usually vertical
C. Is more often used as a receiving antenna
D. Any of the above

3H104
Referred to the fundamental frequency a shorted stub line attached to the transmission line to absorb even harmonics could have a wavelength of:

A. 1.41 wavelength
B. 1/2 wavelengths
C. 1/4 wavelengths
D. 1/6 wavelengths

3H105
Nitrogen gas in concentric Rf transmission lines is used to:

A. Keep moisture out
B. Prevent oxidation
C. Act as insulator
D. Both A and B

3H106
"Stacking" elements on an antenna:

A. Makes for better reception
B. Makes for poorer reception
C. Decreases antenna current
D. Decreases directivity

3H107
The following figure shows the radiated:

A. Voltage along a straightened out balanced loop
B. Voltage along a 1/4 wave Hertz antenna
C. Voltage along a 1/2 wave Hertz antenna
D. Current along a 1/2 wave Hertz antenna

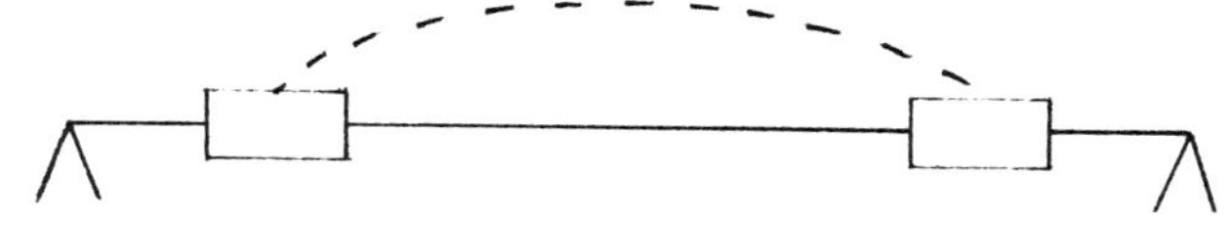

3H108
The parasitic elements on a receiving antenna:

A. Increase its directivity
B. Decrease its directivity
C. Have no effect on its impedance
D. Make it more nearly omnidirectional

3H109
Antenna voltage is:

A. Inversely proportional to the square of its length
B. Proportional to its effective height
C. Measured voltage times length in feet
D. Always proportional to field strength

3H110
The resonant frequency of a Hertz antenna can be lowered by:

A. Lowering the frequency of the transmitter
B. Placing a condenser in series with the antenna
C. Placing a resistor in series with the antenna
D. Placing an inductance in series with the anten:

3H111
Parasitic elements are useful in a receiving antenna for:

A. Increasing directivity
B. Increasing selectivity
C. Increasing sensitivity
D. Both a and c

3H112
This diagram most nearly indicates the:

A. Voltage along a 1/2 wave Hertz antenna
B. Current along a 1/2 wave Hertz antenna
C. Voltage along a 1/4 wave Marconi antenna
D. Current along a 1/4 wave Marconi antenna

3H113
In regards to shipboard satellite dish antenna systems, azimuth is:

A. Vertical aiming of the antenna
B. Horizontal aiming of the antenna
C. 0 - 90 degrees
D. North to east

3H114
What is the effect of adding a capacitor in series to an antenna?

A. Resonant frequency will decrease
B. Resonant frequency will increase
C. Resonant frequency will remain same
D. Electrical length will be longer

3H115
If a transmission line has a power loss of 6 dB per 100 feet, what is the power at the feed point to the antenna at the end of a 200 foot transmission line fed by a 100 watt transmitter?

A. 70 watts
B. 50 watts
C. 25 watts
D. 6 watts

3H116
Waveguides are:

A. Used exclusively in high frequency power supplies
B. Ceramic couplers attached to antenna terminals
C. High-pass filters used at low radio frequencies
D. Hollow metal conductors used to carry high frequency current

3H117
Which of the following represents the best SWR?

A. 1:1
B. 1:2
C. 1:15
D. 2:1

3H118
A 520 kHz signal is fed to a 1/2 wave Hertz antenna. The fifth harmonic will be:

A. 2.65 MHz.
B. 2650 kHz.
C. 2600 kHz.
D. 104 kHz.

3H119
When a capacitor is connected in series with a Marconi antenna:

A. An inductor of equal value must be added
B. No change occurs to antenna
C. Antenna open circuit stops transmission
D. Antenna resonant frequency increases

3H120
How do you increase the electrical length of an antenna?

A. Add an inductor in parallel
B. Add an inductor in series
C. Add a capacitor in series
D. Add a resistor in series

3H121
A coaxial cable has 7 dB of reflected power when the input is 5 watts. What is the output of the transmission line?

A. 5 watts
B. 2.5 watts
C. 1.25 watts
D. 1 watt

3H122
What is the 7th harmonic of 450 kHz when fed through a 1/4 wavelength vertical antenna?

A. 3150 Hz
B. 3150 MHz
C. 787.5 kHz
D. 3.15 MHz

3H123
What is the 5th harmonic of a 450 kHz. transmitter carrier fed to a 1/4 wave antenna?

A. 562.5 MHz
B. 1125 kHz
C. 2250 MHz
D. 2.25 MHz

3H124
Waveguide construction:

A. Should not use silver plating
B. Should not use copper
C. Should have short vertical runs
D. Should not have long horizontal runs

3H125
This drawing indicates:

A. Current along a 1/2 wave Marconi antenna
B. Current along a 1/4 wave Hertz antenna
C. Voltage along a 1/2 wave Hertz antenna
D. Voltage along a 1/4 wave Hertz antenna

3H126
To lengthen an antenna electrically, add a:

A. Coil
B. Resistor
C. Battery
D. Conduit

3H127
How do you electrically decrease the length of an antenna?

A. Add an inductor in series
B. Add a capacitor in series
C. Add an inductor in parallel
D. Add a resistor in series

3H128
If the length of an antenna is changed from 1.5 feet to 1.6 feet its resonant frequency will:

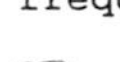

A. Decrease
B. Increase
C. Be 6.7% higher
D. Be 6% lower

3H129
To couple energy into and out of a waveguide:

A. Use wide copper sheeting
B. Use an LC circuit
C. Use capacitive coupling
D. Use a thin piece of wire as an antenna

3H130
An isolator:

A. Acts as a buffer between a microwave oscillator coupled to a waveguide
B. Acts as a buffer to protect a microwave oscillator from variations in line load changes
C. Shields UHF circuits from RF transfer
D. Both A and B

3H131
A high SWR creates losses in transmission lines. A high standing wave ratio might be caused by:

A. Improper turns ratio between primary and secondary in the plate tank transformer
B. Screen grid current flow
C. An antenna electrically too long for its frequency
D. An impedance mismatch

3H132
A properly installed shunt-fed, 1/4 wave marconi antenna:

A. Has zero resistance to ground
B. Has high resistance to ground
C. Should be cut to 1/2 wave
D. Should not be shunt-fed

3H133
When a capacitor is connected in series with a marconi antenna:

A. An inductor of equal value must be added
B. No change occurs to antenna
C. Antenna open circuit stops transmission
D. Antenna resonant frequency increases

3H134
When excited by RF, a half-wave antenna will radiate:

A. A space wave
B. A ground wave
C. Electomagnetic fields
D. Both electromagnetic and electrostatic fields

3H135
Wave guides are:

A. Used exclusively in high frequency power supplies
B. Ceramic couplers attached to anatenna terminals
C. High pass filters used at low radio frequencies
D. Hollow metal conductors used to carry high frequency current

3H136
A 520 kHz signal is fed to a 1/2 wave hertz antenna. The fifth harmonic will be:

A. 2.65 MHz
B. 2650 kHz
C. 2600 kHz
D. 1300 kHz

3H137
The voltage produced in a receiving antenna is:

A. Out of phase with the current if connected properly
B. Out of phase with the current if cut to 1/3 wavelength
C. Variable depending on the station's SWR
D. Always proportional to the received field strength

3H138
A properly connected transmission line:

A. Is grounded at the transmitter end
B. Is cut to a harmonic of the carrier frequency
C. Is cut to an even harmonic of the carrier frequency
D. Has a SWR as near as 1:1 as possible

3H139
Conductance takes place in a waveguide:

A. By interelectron delay
B. Through electrostatic field reluctance
C. In the same manner as a transmission line
D. Through electromagnetic and electrostaatic fields in the walls of the waveguide

3H140
In regards to shunt-fed 1/4 wavelength marconi antenna:

A. DC resistance of the antenna to ground is zero
B. RF resistance from antenna feed point to ground is zero
C. Harmonic radiation is zero under all conditions
D. It must be grounded at both feed and far ends

3H141
What does this drawing indicate?

A. Current along a 1/2 wave marconi antenna
B. Current along a 1/4 wave Hertz antenna
C. Voltage along a 1/2 wave Hertz antenna
D. Voltage along a 1/4 wave hertz antenna

3H142
If a 3/4 wavelength transmission is shortened at one end, impedance at the open will be:

A. Zero
B. Infinite
C. Decreased
D. Increased

3H143
A dummy antenna is a:

A. Non-directional receiver antenna
B. Wide bandwidth directional receiver antenna
C. Transmitter test antenna designed for minimum radiation
D. Transmitter non-directional narrow-band antenna

Radar Questions Element 8

At the time of publication, the FCC has placed a temporary limitation on applicants applying to take the Element #8 examination. Currently, you must have a legal requirement to qualify to take the Radar exam. Until the COLEMS are given authority to offer Radar examinations, you must obtain a letter from your current or future employer - or satisfy the FCC that you are going into business for yourself and will be operating, adjusting and repairing FCC Marine and Avionics transmitters, radar equipment, etc.

Your local FCC Field Office will be conducting the Element #8 radar examination for specific individuals that need to obtain the license for immediate job related requirements (until COLEMS are given the authority to administer tests).

Anyone not needing a license for immediate employment or for licensed repairs, etc., will not be allowed to take the test. If you have any questions that cannot be answered by your local FCC Field Office or COLEM, you may call the headquarters of the FCC in Washington, DC., at (202) 632-4964 - or write to them: Federal Communications Commission, 1919 "M" Street, N.W., Washington, DC. 20554.

The following questions cover Element #8 Radar Endorsement exams. These questions are in a simple question/answer format - without the the three wrong answers. When additional tests become available, they will be published in supplement form to accompany this guidebook:

When handling silicon crystal rectifier cartridges, care should be taken to: Discharge any static charge before touching the crystal.

When may a Ship Radar station be operated by personnel without an FCC license? If no internal adjustments are necessary to operate the pulse magnetron.

What determines the timing duration and shape of the radar pulse? Artificial transmission line.

Range markings are produced on a PPI radar scope by: Applied pulses to the grid of the CRT.

When listening to a radiotelephone receiver, a high level of "hash" noise is usually caused by: radar motor generator or poor grounding or shielding.

A radar installation has a maximum range of 20 miles. The radar beam will travel this distance to a target and bounce back to the receiving antenna in approximately: 250 microseconds.

What is the steady-tone heard on radiotelephone receivers? Radar transmitter pulse or harmonic.

A band of frequencies that ship radar operate: 9300 to 9500 MHz.

What determines the operating frequency of a self-blocking oscillator? Grid circuit time constant.

What precautions should be taken to prevent damage to the magnet of a magnetron? Excessive heat, shocks or magnetized metal filings.

A TR Box has: Two electrodes is a resonant cavity with a spark gap.

Target information on radar is displayed: Cathode-ray tube screen.

What type of modulation causes the radar transmitter carrier to be turned on and off a regular intervals? Pulse.

What frequencies are used for IF amplifiers in radar sets? 30 or 60 MHz.

When adjusting a radar set with an echo box, the serviceman adjusts: For longest "ringing time" and longest "spokes" on radar screen.

What portion of the magnetron is operated at negative potential? Anode.

The rotation position of the radar trace line is determined by: Rotation of deflection coil.

What is the duty cycle of a 1.0 microsecond radar pulse if the repetition rate is 1000 and the average power is 20 watts? 0.001.

What type of transmission line is used most often after the diplexer circuit? Waveguide.

What should be checked when all targets look unusually weak on a radar screen? Mixer crystal.

When installing waveguides, it is important to avoid: Lengthy horizontal sections.

TR and anti-TR tubes act as what type of circuits during transmission? Short Circuits.

What simulates an artificial target utilized to test a radar receiver? Echo box.

What components may be replaced in ship radar equipment by an unlicensed operator? Fuses and receiving tubes.

What is a typical front-to-back resistance ratio for a radar receiver mixer crystal? 20:1

Sea Return on a radar scope is caused by: the reflection of the radar beam bouncing off waves of the sea near the ship.

A waveguide can be terminated at the radar antenna reflector by: A Horn radiator or Parabolic reflector.

Who makes entries in the installation and maintenance log of a ship radar station? Person responsible for operator activity.

Interference to commercial radiotelephone receivers is: Louder on some frequencies than others.

When working with crystals, to avoid possible damage always: Avoid static discharge to the crystal.

What is the peak power of a radar installation having a pulse width of 1.0 microsecond, average power of 18 watts, and the pulse repetition rate of 900? 20,000 watts.

Radar interference to a communications receiver is sometimes characterized by: A steady tone in the headphones.

Bright flashing pie section on a radar PPI scope may be caused by: Defective crystal in the radar receiver.

What is the average power of a 1.0 microsecond radar pulse, when the pulse repetition rate is 900 and the peak power is 25,000 watts? 25 watts.

When do anti-TR and TR tubes deionize? During the receiving interval when the transmitter is inactive (no transmission).

When there are sweep and range marks with no targets and low magnetron current, this indicates: Defective Magnetron.

Flanges and waveguide components such as a parabolic reflector may be physically separated by: 3 millimeters.

Interference to radio equipment caused by a radar set can sometimes be cleared up by: Grounding shields on cables interconnecting the component radar units.

To reduce condensed moisture inside a waveguide: Drill a small 1/8th inch hole at the lowest point of the waveguide.

The radar serviceman who makes an initial radar installation on a ship: Should enter in the log the date and place of installation.

A person whose operator license does not contain a ship radar endorsement may: Replace burned out fuses in the radar set.

Before making repairs or adjustments to ship radar equipment the maintenance person should first: shut off all power.

When a radar set has been initially installed or adjusted and is ready to go into regular service, it is standard policy for the serviceman to: Check other radio and electronic equipment for interference.

Waveguides are used in preference to coaxial lines for the transmission of microwave energy primarily because waveguides: Have less attenuation.

The frequency of a reflex klystron is normally controlled by adjusting: The repeller voltage and the cavity.

A magnetron is: A diode.

What forms the burst of microwave RF energy which is fed to a radar antenna? Magnetron.

The echo box in a radar unit is: A unit for checking radar performance when no targets are available.

In order to prevent weakening of the magnetic field in a magnetron, service operators should: Keep screwdrivers, pliers, etc., from contact with its pole pieces.

A radar receiver mixer crystal is defective when an ohmeter test indicates the resistance is: The same in both directions through the crystal.

Crystal mixers for radar receivers are stored in metal containers to: Minimize temperature changes.

One of the frequency bands used by shipboard radar is: 2900-3100 MHz.

The amount of sea return on a radar scope is controlled by the: Sensitivity time-control circuit.

How does radar interference appear on a LORAN? Narrow vertical "spikes" moving across the screen.

A narrow pulse form a radar transmitter is desirable because: The pulse must end before the reflection can be properly received.

A radar transmitter has a pulse width of 1 microsecond, a peak power of 25 KW., and a pulse repetition of 1000 Hz. Theoretically, what is the closest possible range than can be indicated on the radar scope? 492 feet.

If a radar set has an average power output of 30 watts, pulse duration of 1 microsecond and a repetition rate of 1000: The duty cycle is 0.001.

Aquadag coating on cathode ray tubes are commonly used to: Eliminate effects of external magnetic fields.

The maximum distance a radar system will detect a target is determined by: Peak power.

"Heading Flash" on shipboard radar is: A bright line which appears on the PPI scope to indicate the heading of the ship.

What is utilized as a local oscillator of a radar receiver? Reflex Klystron.

Range marking signals are applied to: Grid circuit of CRT.

A radar PPI cathode ray tube should not be removed from its shipping carton until immediately prior to installing in the radar receiver because: The tube's fluorescent coating is poisonous and may cause injury to maintenance persons.

The FCC requires waveguides installed between the transmitter and antenna to be: Minimum length as practical.

Pulse repetition rate of shipboard radar is: Dependent on the effective range of the radar installation.

What type of presentation is usually used in marine radar equipment? The plan position indicator.

To avoid high voltage shocks and to reduce insulation problems, the magnetron should be: At ground potential.

Range marker rings on a PPI screen are produced by: Electric pulses from a timing circuit.

What is the length of a waveguide antenna stub of a ship radar installation? 1/4 wavelength.

The Discriminator stage in a radar set is tuned to the frequency of what device? Klystron.

A PPI cathode ray tube can be damaged by: Increasing the intensity of the beam and permitting it to remain stationary on the radar scope.

When handling cathode ray tubes, the serviceman should: Wear gloves and goggles.

What equipment aboard ship is affected by radar interference? Auto-Alarm, direction finder, LORAN, receivers, PA systems, etc.

What type of radar beam is best able to separate targets at the same range but different azimuth? Narrow width.

What is the peak power of a 1.0 microsecond radar pulse, when the pulse repetition rate is 900 and the average power is 14 watts? 15,500 watts.

What is the preferred type of transmission conductor of microwave energy in most ship radar installations? Rectangular shaped waveguides.

A high magnetron current reading and a change in oscillator frequency indicates: Defective magnetron magnet.

What is an advantage of using a waveguide over coaxial cable? It attenuates less RF energy.

What is logged in the radar maintenance log? Description of an interference problem and details about any repairs made.

What indications on a LORAN scope indicate the radar is causing interference? "Grass" or narrow vertical "spike" pulses moving moving across the screen.

What determines the operating frequency of a self-blocking oscillator? Bias capacitor and resistor.

A radar installation has a maximum range of 30 miles. The radar beam will travel to the target and bounce back to the receiving antenna in approximately: 370 microseconds.

What type of voltage is applied to the TR tube electrodes to keep gas in the tube at the point of ionization? Keep-alive voltage.

Who may operate Ship Radar equipment? The Master or any person assigned to the job by the ship's Mate.

To prevent mositure from entering a choke-coupling flange: Tightly join flanges with suitable gasket.

A weak magnetron magnet will cause: High current.

What type of detector is normally used in radar receivers? Crystal diode.

The distance to a target when it takes 123 microseconds for a radar pulse to travel to the target and bounce back to the antenna is: 10 miles.

If a TR box malfunctions during the transmission pulse: Possuble damage to the receiver could occur.

The radar should be installed aboard ship in such a way as to: Avoid the maximum number of scanning obstructions.

How is radar interference located with direction finding equipment? Rotate D/F loop antenna until source is detected.

Radar interference to LORAN is sometimes indicated on the LORAN scope as: Vertical "spikes" on the scope.

When the power is shut off and the radar technician is servicing a radar set, he should be sure to: Discharge high voltage capacitors.

A magnetron should not be pulsed with the magnet in place because: Excessive current will flow and will damage the Magnetron.

The sensitivity time control circuit: lowers the sensitivity of the receiver when receiving nearby signals.

The "Bearing Resolution" of a radar set is the ability to distinguish: Minimum angular distance between two targets at the same range.

To minimize signal loss, what should be done to the interior of a waveguide? Keep it as clean as possible.

An artificial target used to check and align radar sets: Echo box.

Gas diodes which ionize and arc during excitation of the transmitter pulse are: TR and anti-TR boxes.

To reduce interference caused by radar and prevent possibility of shock to operators: Attach all metal components thoroughly to the ship's electrical ground.

One of the frequency bands used by shipboard radar is: 5460-5650 MHz.

DIAGRAMS:

CRYSTAL MIXER

KLYSTRON OSCILLATOR

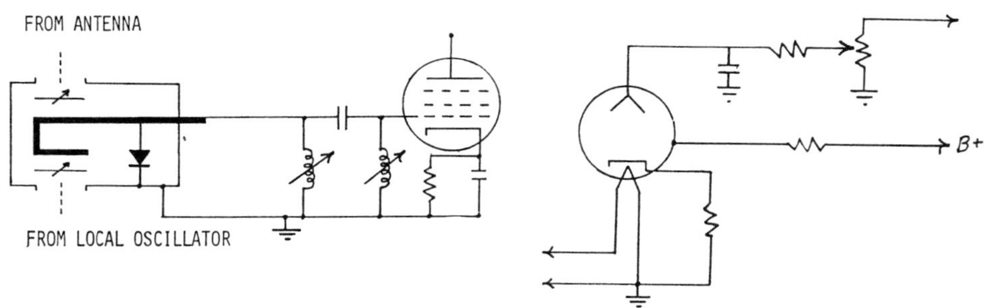

BLOCK DIAGRAM OF RADAR RECEIVER

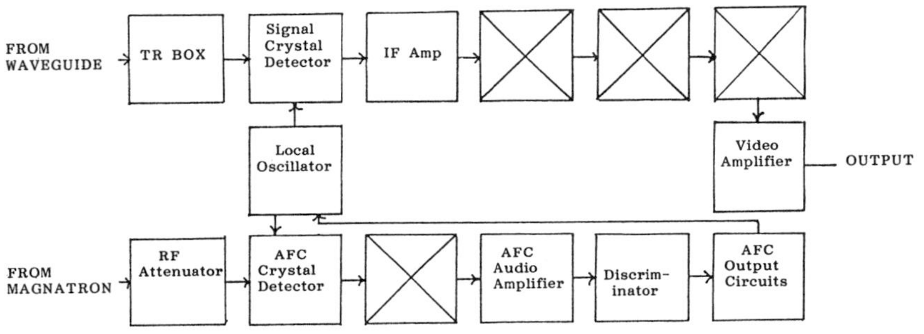

Element 1 Answers

1. A
2. C
3. A
4. B
5. B
6. D
7. A
8. D
9. B
10. C
11. C
12. A
13. C
14. A
15. B
16. C
17. C
18. D
19. D
20. C
21. A
22. C
23. C
24. B
25. D
26. B
27. A
28. D
29. C
30. B
31. C
32. B
33. C
34. A
35. C
36. C
37. A
38. D
39. B
40. B
41. B
42. A
43. D
44. A
45. C
46. A
47. C
48. B
49. B
50. C
51. A
52. B
53. C
54. A
55. B
56. D
57. C
58. B
59. B
60. A
61. B
62. C
63. B
64. A
65. B
66. A
67. C
68. A
69. D
70. D
71. C
72. C
73. A
74. A
75. D
76. A
77. D
78. B
79. D
80. A
81. D
82. B
83. C
84. C
85. D
86. A
87. A
88. D
89. B
90. A
91. B
92. D
93. C
94. B
95. B
96. B
97. D
98. C
99. D
100. D
101. D
102. D
103. B
104. A
105. A
106. A
107. B
108. D
109. C
110. D
111. A
112. D
113. B
114. D
115. C
116. D
117. D
118. B
119. D
120. C
121. C
122. A
123. D
124. C
125. A
126. C
127. C
128. C
129. C
130. B
131. A
132. D
133. C
134. B
135. C
136. C
137. D
138. D
139. A
140. B
141. A
142. D
143. C
144. A
145. D
146. A
147. D
148. C
149. B
150. D
151. B
152. C
153. A
154. C
155. A
156. A
157. D
158. A
159. B
160. D
161. D
162. B
163. A
164. D
165. A
166. D
167. D
168. D
169. C
170. B

Element 3A Answers

1.	D	11.	D	21.	D	31.	D
2.	A	12.	A	22.	A	32.	C
3.	B	13.	B	23.	B	33.	C
4.	B	14.	B	24.	B	34.	B
5.	C	15.	A	25.	A	35.	C
6.	C	16.	C	26.	B	36.	C
7.	A	17.	B	27.	B	37.	A
8.	C	18.	C	28.	C	38.	D
9.	B	19.	D	29.	A	39.	B
10.	A	20.	C	30.	A	40.	D

Element 3B Answers

1.	B	7.	A	13.	C	19.	A
2.	C	8.	B	14.	C	20.	C
3.	D	9.	C	15.	A	21.	C
4.	B	10.	A	16.	C	22.	D
5.	A	11.	B	17.	B		
6.	D	12.	A	18.	C		

Element 3C Answers

1.	B	26.	B	51.	C	76.	C
2.	A	27.	D	52.	D	77.	A
3.	B	28.	B	53.	B	78.	A
4.	A	29.	C	54.	C	79.	D
5.	D	30.	A	55.	C	80.	A
6.	C	31.	C	56.	B	81.	C
7.	A	32.	A	57.	D	82.	C
8.	D	33.	D	58.	C	83.	B
9.	D	34.	B	59.	D	84.	C
10.	A	35.	A	60.	C	85.	A
11.	C	36.	C	61.	D	86.	A
12.	C	37.	C	62.	A	87.	D
13.	D	38.	C	63.	B	88.	A
14.	B	39.	A	64.	D	89.	B
15.	D	40.	C	65.	A	90.	B
16.	B	41.	D	66.	B	91.	D
17.	A	42.	A	67.	A	92.	B
18.	B	43.	A	68.	C	93.	D
19.	A	44.	B	69.	C	94.	D
20.	D	45.	D	70.	A	95.	C
21.	B	46.	C	71.	D	96.	C
22.	D	47.	D	72.	C	97.	A
23.	C	48.	A	73.	D		
24.	D	49.	A	74.	A		
25.	B	50.	A	75.	A		

Element 3D Answers

1.	A	30.	D	59.	B	88.	C
2.	D	31.	B	60.	B	89.	D
3.	A	32.	C	61.	C	90.	B
4.	B	33.	D	62.	C	91.	C
5.	C	34.	A	63.	B	92.	B
6.	B	35.	D	64.	D	93.	C
7.	D	36.	B	65.	A	94.	B
8.	B	37.	C	66.	A	95.	A
9.	A	38.	D	67.	B	96.	D
10.	B	39.	A	68.	A	97.	C
11.	B	40.	D	69.	D	98.	C
12.	A	41.	B	70.	A	99.	C
13.	C	42.	A	71.	C	100.	A
14.	A	43.	D	72.	D	101.	C
15.	A	44.	C	73.	D	102.	D
16.	C	45.	D	74.	A	103.	A
17.	B	46.	A	75.	D	104.	D
18.	D	47.	B	76.	C	105.	C
19.	C	48.	D	77.	A	106.	C
20.	B	49.	D	78.	B	107.	D
21.	B	50.	C	79.	A	108.	C
22.	A	51.	B	80.	D	109.	D
23.	D	52.	A	81.	A	110.	D
24.	A	53.	D	82.	B	111.	D
25.	A	54.	D	83.	C	112.	B
26.	C	55.	C	84.	D	113.	C
27.	D	56.	D	85.	A	114.	C
28.	C	57.	C	86.	A	115.	C
29.	B	58.	A	87.	B		

Element 3E Answers

1.	A	20.	A	39.	A	58.	C
2.	C	21.	B	40.	D	59.	D
3.	B	22.	B	41.	D	60.	A
4.	A	23.	B	42.	B	61.	D
5.	C	24.	B	43.	C	62.	D
6.	C	25.	A	44.	D	63.	B
7.	A	26.	A	45.	D	64.	B
8.	B	27.	C	46.	A	65.	C
9.	D	28.	A	47.	D	66.	A
10.	D	29.	A	48.	A	67.	A
11.	B	30.	B	49.	D	68.	C
12.	D	31.	A	50.	C	69.	D
13.	C	32.	A	51.	A	70.	B
14.	D	33.	A	52.	D	71.	B
15.	C	34.	D	53.	D	72.	C
16.	C	35.	A	54.	A	73.	C
17.	B	36.	B	55.	A	74.	B
18.	C	37.	C	56.	B	75.	A
19.	C	38.	B	57.	A		

Element 3F Answers

1. D
2. C
3. A
4. B
5. D
6. C
7. A
8. D
9. B
10. B
11. A
12. D
13. B
14. A
15. A
16. C
17. C
18. A
19. D
20. B
21. D
22. B
23. D
24. C
25. D
26. B
27. A
28. D
29. C
30. A
31. C
32. A
33. D
34. C
35. B
36. B
37. D
38. D
39. C
40. D
41. D
42. A
43. B
44. B
45. C
46. D
47. B
48. C
49. B
50. D
51. D
52. A
53. B
54. B
55. B
56. C
57. B
58. A
59. C
60. B
61. A
62. D
63. C
64. B
65. A
66. B
67. C
68. C
69. A
70. C
71. D
72. B
73. D
74. A
75. C
76. C
77. C
78. C
79. B
80. D
81. A
82. A
83. D
84. A
85. C
86. A
87. C
88. D
89. A
90. A
91. D
92. D
93. C
94. D
95. A
96. B
97. B
98. D
99. B
100. D
101. C
102. A
103. D
104. A
105. D
106. A
107. B
108. B
109. B
110. D
111. D
112. B
113. B
114. B
115. A
116. D
117. D
118. C
119. D
120. A
121. B
122. C
123. C
124. A
125. B
126. D
127. C
128. A
129. C
130. A
131. C
132. C
133. D
134. D
135. D
136. D
137. D
138. B
139. D

Element 3G Answers

1.	A	26.	C	51.	D	76.	A
2.	B	27.	A	52.	A	77.	B
3.	C	28.	B	53.	C	78.	B
4.	D	29.	C	54.	C	79.	C
5.	A	30.	D	55.	C	80.	C
6.	B	31.	A	56.	B	81.	C
7.	B	32.	B	57.	D	82.	A
8.	D	33.	B	58.	A	83.	A
9.	C	34.	C	59.	A	84.	D
10.	C	35.	D	60.	C	85.	B
11.	D	36.	A	61.	D	86.	D
12.	A	37.	B	62.	A	87.	A
13.	B	38.	C	63.	C	88.	B
14.	C	39.	D	64.	A	89.	D
15.	D	40.	A	65.	B	90.	D
16.	A	41.	B	66.	C	91.	A
17.	B	42.	C	67.	D	92.	B
18.	D	43.	D	68.	B	93.	A
19.	D	44.	A	69.	C	94.	A
20.	A	45.	C	70.	B	95.	D
21.	B	46.	C	71.	B	96.	B
22.	C	47.	D	72.	B	97.	A
23.	D	48.	A	73.	B		
24.	A	49.	B	74.	D		
25.	B	50.	C	75.	A		

Element 3H Answers

1.	A	37.	C	73.	C	109.	B
2.	B	38.	D	74.	C	110.	D
3.	B	39.	B	75.	B	111.	C
4.	C	40.	C	76.	A	112.	C
5.	C	41.	D	77.	D	113.	B
6.	D	42.	A	78.	A	114.	B
7.	A	43.	B	79.	C	115.	D
8.	B	44.	C	80.	B	116.	D
9.	C	45.	D	81.	C	117.	A
10.	D	46.	A	82.	D	118.	C
11.	A	47.	A	83.	A	119.	D
12.	B	48.	D	84.	C	120.	B
13.	C	49.	B	85.	D	121.	D
14.	B	50.	B	86.	A	122.	D
15.	C	51.	A	87.	D	123.	D
16.	B	52.	B	88.	B	124.	D
17.	A	53.	D	89.	A	125.	C
18.	B	54.	D	90.	A	126.	A
19.	C	55.	A	91.	D	127.	B
20.	D	56.	A	92.	B	128.	A
21.	A	57.	B	93.	C	129.	D
22.	B	58.	C	94.	C	130.	D
23.	C	59.	D	95.	B	131.	D
24.	D	60.	B	96.	C	132.	A
25.	A	61.	B	97.	D	133.	D
26.	D	62.	C	98.	C	134.	D
27.	A	63.	D	99.	A	135.	D
28.	B	64.	C	100.	D	136.	C
29.	C	65.	C	101.	C	137.	D
30.	B	66.	B	102.	D	138.	D
31.	B	67.	A	103.	D	139.	D
32.	B	68.	D	104.	C	140.	A
33.	C	69.	C	105.	D	141.	C
34.	A	70.	C	106.	A	142.	B
35.	A	71.	A	107.	D	143.	C
36.	B	72.	B	108.	A		

Part 13; FCC Rules and Regulations

Here is the revised Amendment of Part 13 of the Federal Communications rules regarding administration of examinations for Commercial Operator Licenses. All of the following rules and regulation information is NOT required knowledge to pass FCC License exams. However, these "Part 13" government regulations are reprinted for your general knowledge.

This section of the guidebook may provide you with a better understanding of many FCC operation and procedure test questions. There is no need to memorize all of the information on the following pages.

GENERAL

§ 13.1 Basis and purpose.
§ 13.3 Definitions.
§ 13.5 Licensed commercial radio operators required.
§ 13.7 Classification of operator licenses and endorsements.
§ 13.9 Eligibility and application for new license or endorsement.
§ 13.11 Holding more that one commercial radio operator license.
§ 13.13 Application for renewed or modified license.
§ 13.15 License term.
§ 13.17 Replacement license.
§ 13.19 Operator's responsibility.

EXAMINATION SYSTEM

§ 13.201 Qualifying for a commercial operator license or endorsement.
§ 13.203 Examination elements.
§ 13.205 Examination credit for licenses held.
§ 13.207 Preparing an examination.
§ 13.209 Examination procedures.
§ 13.211 Commercial radio operator license examination.
§ 13.213 COLEM qualifications.
§ 13.215 Question pools.
§ 13.217 Records.

GENERAL PROVISIONS

§ 13.1 Basis and purpose.

(a) Basis. The basis for the rules contained in this Part is the Communications Act of 1934, as amended, and applicable treaties and agreements to which the United States is a party.

(b) Purpose. The purpose of the rules in this Part is to prescribe the manner and conditions under which commercial radio operators are licensed by the Commission.

§ 13.3 Definitions.

The definitions of terms used in Part 13 are:

(a) COLEM. Commercial operator license examination manager.

(b) Commercial radio operator. A person holding a license or licenses specified in § 13.7(b).

(c) GMDSS. Global Maritime Distress and Safety System.

(d) FCC. Federal Communications Commission.

(e) International Morse code. A dot-dash code as defined in International Telegraph and Telephone Consultative Committee (CCITT) Recommendation F.1 (1984), Division B, I. Morse code.

(f) ITU. International Telecommunication Union.

(g) PPC. Proof-of-Passing Certificate.

(h) Question pool. All current examination questions for a designated written examination element.

(i) Question set. A series of examination questions on a given examination selected from the current question pool.

(j) Radio Regulations. The latest ITU Radio Regulations to which the United States is a party.

§ 13.5 Licensed commercial radio operator required.

Rules that require FCC station licensees to have certain transmitter operating, maintenance, and repair duties performed by a commercial radio operator are contained in Parts 23, 73, 74, 80, and 87 of this Chapter.

§ 13.7 Classification of operator licenses and endorsements.

(a) Commercial radio operator licenses issued by the FCC are classified in accordance with the Radio Regulations of the ITU.

(b) There are nine types of commercial radio operator licenses, certificates and permits (licenses). The license's ITU classification, if different from its name, is given in parenthesis.

(1) First Class Radiotelegraph Operator's Certificate.

(2) Second Class Radiotelegraph Operator's Certificate.

(3) Third Class Radiotelegraph Operator's Certificate (radiotelegraph operator's special certificate).

(4) General Radiotelephone Operator License (radiotelephone operator's general certificate).

(5) Marine Radio Operator Permit (radiotelephone operator's restricted certificate).

(6) Restricted Radiotelephone Operator Permit (radiotelephone operator's restricted certificate).

(7) Restricted Radiotelephone Operator Permit-Limited Use (radiotelephone operator's restricted certificate).

(8) GMDSS Radio Operator's License (general operator's certificate).

(9) GMDSS Radio Maintainer's License (technical portion of the first-class radio electronic certificate).

(c) There are six license endorsements affixed by the FCC to provide special authorizations or restrictions. Endorsements may be affixed to the license(s) indicated in parenthesis.

(1) Ship Radar Endorsement (First and Second Class Radiotelegraph Operator's Certificates, General Radiotelephone Operator License, GMDSS Radio Maintainer's License).

(2) Six Months Service Endorsement (First and Second Class Radiotelegraph Operator's License).

(3) Restrictive endorsements relating to physical handicaps, English language or literacy waivers, or other matters (all licenses).

(4) Marine Radio Operator Permits shall bear the following endorsement: This Permit does not authorize the operation of AM, FM or TV broadcast stations.

(5) General Radiotelephone Operator Licenses issued after December 31, 1985, shall bear the following endorsement: This license confers authority to operate licensed radio stations in the Aviation, Marine and International Fixed Public Radio Serviced only. This authority is subject to: any endorsement placed upon this license; FCC orders, rules, and regulations; United States statutes; and the provisions of any treaties to which the United States is a party. This license does not confer any authority to operate broadcast stations. It is not assignable or transferable.

(6) (i) If a person is afflicted with an uncorrected physical handicap which would clearly prevent the performance of all or any part of the duties of a radio operator, under the license for which application is made, at a station under emergency conditions involving the safety of life or property, that person still may be issued the license if found qualified. Such a license shall bear a restrictive endorsement as follows:

This license is not valid for the performance of any operating duties, other than installation, service and maintenance duties, at any station licensed by the FCC which is required, directly or indirectly, by any treaty, statute or rule or regulation pursuant to statute, to be provided for safety purposes.

(ii) In the case of a license that does not require an examination in technical radio matters, the endorsement specified in (i) above will be modified by deleting the reference therein to installation, service, and maintenance duties.

(iii) In any case where an applicant who normally would receive or has received a commercial radio operator license bearing the endorsement prescribed by paragraph (i) above, indicates a desire to operate a station falling within the prohibited terms of the endorsement, the applicant may request in writing that such endorsement not be placed upon, or be removed from his or her license, and may submit written comments or statements from other parties in support thereof.

(iv) An applicant who shows that he has performed satisfactorily the duties of a radio operator at a station required to be provided for safety purposes during a period when he or she was afflicted by uncorrected physical handicaps of the same kind and to the same degree as the physical handicaps shown by his or her current application shall not be deemed to be within the provisions of paragraph (i) above.

(d) A Restricted Radiotelephone Operator Permit-Limited Use issued by the FCC to an aircraft pilot who is not legally eligible for employment in the United States is valid only for operating radio stations on aircraft.

(e) A Restricted Radiotelephone Operator Permit-Limited Use issued by the FCC to a person under the provision of § 303(l)(2) of the Communications Act of 1934, as amended, is valid only for the operation of radio stations for which that person is the station licensee.

§ 13.9 Eligibility and application for new license or endorsement.

(a) If found qualified, the following persons are eligible to apply for commercial radio operator licenses:

(1) Any person legally eligible for employment in the United States.

(2) Any person, for the purpose of operating aircraft radio stations, who holds:

(i) United States pilot certificates; or

(ii) Foreign aircraft pilot certificates which are valid in the United States, if the foreign government involved has entered into a reciprocal agreement under which such foreign government does not impose any similar requirement relating to eligibility for employment upon United States citizens.

(3) Any person who holds a FCC radio station license, for the purpose of operating that station.

(4) Notwithstanding any other provisions of the FCC's rules, no person shall be eligible to be issued a commercial radio operator license when

(i) the person's commercial radio operator license is suspended, or

(ii) the person's commercial radio operator license is the subject of an ongoing suspension proceeding, or

(iii) the person is afflicted with complete deafness or complete muteness or complete inability for any other reason to transmit correctly and to receive correctly by telephone spoken messages in English.

(b) (1) Each application for a new General Radiotelephone Operator License, Marine Radio Operator Permit, First Class Radiotelegraph Operator's Certificate, Second Class Radiotelegraph Operator's Certificate, Third Class Radiotelegraph Operator's Certificate, Ship Radar Endorsement, Six Months Service Endorsement, GMDSS Radio Operator's License or GMDSS Radio Maintainer's License must be made on FCC Form 756.

(2) Each application for a Restricted Radiotelephone Operator Permit must be made on FCC Form 753.

(3) Each application for a Restricted Radiotelephone Operator Permit-Limited Use must be made on FCC Form 755.

(c) Each application for a new General Radiotelephone Operator License, Marine Radio Operator Permit, First Class Radiotelegraph Operator's Certificate, Second Class Radiotelegraph Operator's Certificate, Third Class Radiotelegraph Operator's Certificate, Ship Radar Endorsement, GMDSS Radio Operator's License or GMDSS Radio Maintainer's License must include an original PPC(s) from a COLEM(s) showing that the applicant has passed the necessary examination element(s). The applicant must submit the application to the address specified in the Part 1 of the rules.

(d) Each application for a new six month radiotelegraph endorsement must include documentation showing that the applicant has satisfied the requirements of Section 13.201(c). The applicant must submit the application to the address specified in the Part 1 of the rules.

(e) No person shall alter, duplicate for fraudulent purposes, or fraudulently obtain or attempt to obtain an operator license. No person shall use a license issued to another or a license that he or she knows to be altered, duplicated for fraudulent purposes, or fraudulently obtained. No person shall obtain or attempt to obtain, or assist another person to obtain or attempt to obtain, an operator license by fraudulent means.

§ 13.11 Holding more than one commercial radio operator license.

(a) An eligible person may hold more than one commercial operator license except as follows:

(1) No person may hold two or more unexpired radiotelegraph operator's certificates at the same time;

(2) No person may hold any class of radiotelegraph operator's certificate and a Marine Radio Operator Permit;

(3) No person may hold any class of radiotelegraph operator's certificate and a Restricted Radiotelephone Operator Permit.

(b) Each person who is not legally eligible for employment in the United States, and certain other persons who were issued permits prior to September 13, 1982, may hold two Restricted Radiotelephone Operator Permits simultaneously when each permit authorizes the operation of a particular station or class of stations.

§ 13.13 Application for a renewed or modified license.

(a) Each application to renew a First Class Radiotelegraph Operator's Certificate, Second Class Radiotelegraph Operator's Certificate, Third Class Radiotelegraph Operator's Certificate, GMDSS Radio Operator's License, or GMDSS Radio Maintainer's License must be made on FCC Form 756. The application must be accompanied by the original document or a legible photocopy unless it has been lost, mutilated, or destroyed. If the license has been lost, mutilated, or destroyed, submit a written explanation. The application must be accompanied by the appropriate fee and submitted to the address specified in Part 1 of the rules.

(b) A licensee may submit an application for renewal of an unexpired license during the last year of the license term. If a license expires, application for renewal may be made during a grace period of five years after the expiration date without having to retake the required examinations. The application must be accompanied by the required fee and submitted to the address specified in Part 1 of the rules. During the grace period, the expired license is not valid. A license renewed during the grace period will be effective as of the date of the renewal. Licensees who fail to renew their license within the grace period must apply for a new license and take the required examination(s).

(c) Each application involving a change in operator class must be made on FCC Form 756. Each application for a commercial operator license involving a change in operator class must include original PPC(s) from a COLEM showing that the applicant has passed the necessary examination element(s). The application must be accompanied by the required fee, if any, and submitted to the address specified in Part 1 of the rules.

(d) The holder of a First Class Radiotelegraph Operator's Certificate, Second Class Radiotelegraph Operator's Certificate, Third Class Radiotelegraph Operator's Certificate, General Radiotelephone Operator License, GMDSS Radio Operator's License, or GMDSS Radio Maintainer's License whose name is legally changed may obtain a modified license by filing a FCC Form 756 with a written explanation. The application must be accompanied by the required fee and submitted to the address specified in Part 1 of the rules.

(e) A licensee who has made application for a renewed or modified operator license or permit may exhibit a photocopy of their license in lieu of the original document.

§ 13.15 License Term.

(a) Commercial radio operator licenses are normally valid for a term of five years from the date of issuance, except as provided in paragraph (b) of this section.

(b) General Radiotelephone Operator Licenses, Restricted Radiotelephone Operator Permits, and Restricted Radiotelephone Operator Permits-Limited Use are normally valid for the lifetime of the holder. The terms of all Restricted Radiotelephone Operator Permits issued prior to November 15, 1953, and valid on that date, are extended to the lifetime of the operator.

§ 13.17 Replacement license.

(a) Each licensee or permittee whose original document is lost, mutilated, or destroyed must request a replacement. The application must be accompanied by the required fee and submitted to the address specified in Part 1 of the rules.

(b) Each application for a replacement General Radiotelephone Operator License, Marine Radio Operator Permit, First Class Radiotelegraph Operator's Certificate, Second Class Radiotelegraph Operator's Certificate, Third Class Radiotelegraph Operator's Certificate, GMDSS Radio Operator's License, GMDSS Radio Maintainer's License, must be made on FCC Form 756 and must include a written explanation as to the circumstances involved in the loss, mutilation, or destruction of the original document.

(c) Each application for a replacement Restricted Radiotelephone Operator Permit must be on FCC Form 753.

(d) Each application for a replacement Restricted Radiotelephone Operator Permit-Limited Use must be on FCC Form 755.

(e) A licensee who has made application for a replacement license may exhibit a copy of the application submitted to the FCC or a photocopy of the license in lieu of the original document.

§ 13.19 Operator's responsibility.

(a) The operator responsible for maintenance of a transmitter may permit other persons to adjust that transmitter in the operator's presence for the purpose of carrying out tests or making adjustments requiring specialized knowledge or skill, provided that he or she shall not be relieved thereby from responsibility for the proper operation of the equipment.

(b) In every case where a station operating log or service and maintenance log is required, the operator responsible for the station operation or maintenance shall make the required entries in the station log. If no station log is required, the operator responsible for service or maintenance duties which may affect the proper operation of the station shall sign and date an entry in the station maintenance records giving:

(1) Pertinent details of all service and maintenance work performed by the operator or conducted under his or her supervision;

(2) His or her name and address; and

(3) The class, serial number and expiration date of the license:

(c) When the operator is on duty and in charge of transmitting systems, or performing service, maintenance or inspection functions, the license or permit document, or a photocopy thereof, must be posted or in the operator's personal possession, and available for inspection upon request by a FCC representative.

(d) The operator on duty and in charge of transmitting systems, or performing service, maintenance or inspection functions, shall not be subject to the requirements of paragraph (b) of this section at a station, or stations of one licensee at a single location, at which the operator is regularly employed and at which his or her license, or a photocopy, is posted.

EXAMINATION SYSTEM

§ 13.201 Qualifying for a commercial operator license or endorsement.

(a) To be qualified to hold any commercial radio operator license, an applicant must have a satisfactory knowledge of FCC rules and must have the ability to send correctly and receive correctly spoken messages in the English language.

(b) An applicant must pass an examination for the issuance of a new commercial radio operator license, other than the Restricted Radiotelephone Operator Permit and the Restricted Radiotelephone Operator Permit-Limited Use, and for each change in operator class. An applicant must pass an examination for the issuance of a new Ship Radar Endorsement. Each applicant for the class of license or endorsement specified below must pass, or otherwise receive credit for, the corresponding examination elements:

(1) First Class Radiotelegraph Operator's Certificate.

(i) Telegraphy Elements 3 and 4;

(ii) Written Elements 1, 5, and 6;

(iii) Applicant must be at least 21 years old;

(iv) Applicant must have one year of experience in sending and receiving public correspondence by radiotelegraph at a public coast station, a ship station, or both.

(2) Second Class Radiotelegraph Operator's Certificate.

(i) Telegraphy Elements 1 and 2;

(ii) Written Elements 1, 5, and 6.

(3) Third Class Radiotelegraph Operator's Certificate.

(i) Telegraphy Elements 1 and 2;

(ii) Written Elements 1 and 5.

(4) General Radiotelephone Operator License: Written Elements 1 and 3.

(5) Marine Radio Operator Permit: Written Element 1.

(6) GMDSS Radio Operator's License: Written Elements 1 and 7.

(7) GMDSS Radio Maintainer's License: Written Elements 1, 3, and 9.

(8) Ship Radar Endorsement: Written Element 8.

(c) An applicant for the six months service endorsement must show that:

(1) The applicant was employed as a radio operator on board a ship or ships of the United States for a period totaling at least six months;

(2) The ships were equipped with a radio station complying with the provisions of Part II of Title III of the Communications Act, or the ships were owned and operated by the U.S. Government and equipped with radio stations;

(3) The ships were in service during the applicable six month period and no portion of any single in-port period included in the qualifying six months period exceeded seven days;

(4) The applicant held a FCC-issued First or Second Class Radiotelegraph Operator's Certificate during this entire six month qualifying period; and

(5) The applicant holds a radio officer's license issued by the U.S.

Coast Guard at the time the six month endorsement is requested.

§ 13.203 Examination elements.

(a) A written examination (written Element) must prove that the examinee possesses the operational and technical qualifications to perform the duties required by a person holding that class of commercial radio operator license. Each written examination must be comprised of a question set as follows:

(1) Element 1 (formerly Elements 1 and 2): Basic radio law and operating practice with which every maritime radio operator should be familiar. 24 questions concerning provisions of laws, treaties, regulations, and operating procedures and practices generally followed or required in communicating by means of radiotelephone stations. The minimum passing score is 18 questions answered correctly.

(2) Element 3: General radiotelephone. 76 questions concerning electronic fundamentals and techniques required to adjust, repair, and maintain radio transmitters and receivers at stations licensed by the FCC in the aviation, maritime, and international fixed public radio services. The minimum passing score is 57 questions answered correctly.

(3) Element 5: Radiotelegraph operating practice. 50 questions concerning radio operating procedures and practices generally followed or required in communicating by means of radiotelegraph stations primarily other than in the maritime mobile services of public correspondence. The minimum passing score is 38 questions answered correctly.

(4) Element 6: Advanced radiotelegraph. 100 questions concerning technical, legal and other matters applicable to the operation of all classes of radiotelegraph stations, including operating procedures and practices in the maritime mobile services of public correspondence, and associated matters such as radio navigational aids, message traffic routing and accounting, etc. The minimum passing score is 75 questions answered correctly.

(5) Element 7: GMDSS radio operating practices. 76 questions concerning GMDSS radio operating procedures and practices sufficient to show detailed practical knowledge of the operation of all GMDSS sub-systems and equipment; ability to send and receive correctly by radio telephone and narrow-band direct-printing telegraphy; detailed knowledge of the regulations applying to radio communications, knowledge of the documents relating to charges for radio communications and knowledge of those provisions of the International Convention for the Safety of Life at Sea which relate to radio; sufficient knowledge of English to be able to express oneself satisfactorily both orally and in writing; knowledge of and ability to perform each function listed in Section 80.1081; and knowledge covering the requirements set forth in IMO Assembly Resolution on Training for Radio Personnel (GMDSS), Annex 3. The minimum passing score is 57 questions answered correctly.

(6) Element 8: Ship radar techniques. 50 questions concerning specialized theory and practice applicable to the proper installation, servicing and maintenance of ship radar equipment in general use for marine navigational purposes. The minimum passing score is 38 questions answered correctly.

(7) Element 9: GMDSS radio maintenance practices and procedures. 50 questions concerning the requirements set forth in IMO Assembly on Training for Radio Personnel (GMDSS), Annex 5 and IMO Assembly on Radio Maintenance Guidelines for the Global Maritime Distress and Safety System related to Sea Areas A3 and A4. The minimum passing score is 38 questions answered correctly.

(b) A telegraphy examination (telegraphy Elements) must prove that the examinee has the ability to send correctly by hand and to receive correctly by ear texts in the international Morse code at not less than the prescribed speed, using all the letters of the alphabet, numerals 0-9, period, comma, question mark, slant mark, and prosigns AR, BT, and SK.

(1) Telegraphy Element 1: 16 code groups per minute.

(2) Telegraphy Element 2: 20 words per minute.

(3) Telegraphy Element 3: 20 code groups per minute.

(4) Telegraphy Element 4: 25 words per minute.

§ 13.205 Examination credit for licenses held.

(a) The COLEM must give credit as specified below to an examinee holding any of the following documents:

(1) An unexpired (or within the grace period) FCC-issued commercial radio operator license: the written examination and telegraphy Element(s) required to obtain the license held.

(2) A PPC: Each element the PPC indicates the examinee passed within the previous 365 days.

(3) An unexpired or within the grace period FCC-issued Amateur Extra Class operator license: Telegraphy Elements 1 and 2.

(b) No examination credit, except as herein provided, shall be allowed on the basis of holding or having held any other license, permit, or certificate.

§ 13.207 Preparing an examination.

(a) Each telegraphy message and each written question set administered to an examinee for a commercial radio operator license must be provided by a COLEM.

(b) Each question set administered to an examinee must utilize questions taken from the applicable Element question pool. The COLEM may obtain the written question sets from a supplier or other COLEM.

(c) A telegraphy examination must consist of a plain language text or code group message sent in the international Morse code at no less than the prescribed speed for a minimum of five minutes. The message must contain each required telegraphy character at least once. No message known to the examinee may be administered in a telegraphy examination. Each five letters of the alphabet must be counted as one word or one code group. Each numeral, punctuation mark, and prosign must be counted as two letters of the alphabet. The COLEM may obtain the telegraphy message from a supplier or other COLEM.

§ 13.209 Examination procedures.

(a) Each examination for a commercial radio operator license must be administered at a location and a time specified by the COLEM. The COLEM is responsible for the proper conduct and necessary supervision of each examination. The COLEM must immediately terminate the examination upon failure of the examinee to comply with its instructions.

(b) Each examinee, when taking an examination for a commercial radio operator license, shall comply with the instructions of the COLEM.

(c) No examination that has been compromised shall be administered to any examinee. Neither the same telegraphy message nor the same question set may be re-administered to the same examinee.

(d) Passing a telegraphy examination.

(1) To pass a receiving telegraphy examination, an examinee is required to receive correctly the message by ear, for a period of 1 minute without error at the rate of speed specified in Section 13.203 for the class of license sought.

(2) To pass a sending telegraphy examination, an examinee is required to send correctly for a period of 1 minute at the rate of speed prescribed in Section 13.203(b) for the class of license sought.

(e) Passing a telegraphy receiving examination is adequate proof of an examinee's ability to both send and receive telegraphy. The COLEM, however, may also include a sending segment in a telegraphy examination.

(f) The COLEM is responsible for determining the correctness of the examinee's answers. When the examinee does not score a passing grade on an examination element, the COLEM must inform the examinee of the grade.

(g) When the examinee is credited for all examination elements required for the commercial operator license sought, the examinee may apply to the FCC for the license.

(h) No applicant who is eligible to apply for any commercial radio operator license shall, by reason of any physical handicap, be denied the privilege of applying and being permitted to attempt to prove his or her qualifications (by examination if examination is required) for such commercial radio operator license in accordance with procedures established by the COLEM.

(i) The COLEM must accommodate an examinee whose physical disabilities require a special examination procedure. The COLEM may require a physician's certification indicating the nature of the disability before determining which, if any, special procedures are appropriate to use. In the case of a blind examinee, the examination questions may be read aloud and the examinee may answer orally. A blind examinee wishing to use this procedure must make arrangements with the COLEM prior to the date the examination is desired.

(j) The FCC may:

(1) Administer any examination element itself.

(2) Readminister any examination element previously administered by a COLEM, either itself or by designating another COLEM to readminister the examination element.

(3) Cancel the commercial operator license(s) of any licensee who fails to appear for re-administration of an examination when directed by the FCC, or who fails any required element that is re-administered. In case of such cancellation, the person will be issued an operator license consistent with completed examination elements that have not been invalidated by not appearing for, or by failing, the examination upon re-administration.

§ 13.211 Commercial radio operator license examination.

(a) Each session where an examination for a commercial radio operator license is administered must be managed by a COLEM or the FCC.

(b) Each examination for a commercial radio operator license must be administered as determined by the COLEM.

(c) The COLEM may limit the number of candidates at any examination.

(d) The COLEM may prohibit from the examination area items the COLEM determines could compromise the integrity of an examination or distract examinees.

(e) Within 10 days of completion of the examination element(s), the COLEM must provide the results of the examination to the examinee and the COLEM must issue a PPC to an examinee who scores a passing grade on an examination element.

(f) A PPC is valid for 365 days from the date it is issued.

§ 13.213 COLEM qualifications.

No entity may serve as a COLEM unless it has entered into a written agreement with the FCC. In order to be eligible to be a COLEM, the entity must:

(a) Agree to abide by the terms of the agreement;

(b) Be capable of serving as a COLEM;

(c) Agree to coordinate examinations for one or more types of commercial radio operator licenses and/or endorsements;

(d) Agree to assure that, for any examination, every examinee eligible under these rules is registered without regard to race, sex, religion, national origin or membership (or lack thereof) in any organization;

(e) Agree to make any examination records available to the FCC, upon request.

(f) Agree not to administer an examination to an employee, relative, or relative of an employee.

§ 13.215 Question pools.

The question pool for each written examination element will be composed of questions acceptable to the FCC. Each question pool must contain at least 5 times the number of questions required for a single examination. The FCC will issue public announcements detailing the questions in the pool for each element. COLEMs must use only the most recent question pool made available to the public when preparing a question set for a written examination element.

§ 13.217 Records.

Each COLEM recovering fees from examinees must maintain records of expenses and revenues, frequency of examinations administered, and examination pass rates. Records must cover the period from January 1 to December 31 of the preceding year and must be submitted as directed by the Commission. Each COLEM must retain records for 1 year and the records must be made available to the FCC upon request.

FCC Field Offices

ALASKA, Anchorage Office
Federal Communications Commission
6721 Raspberry Road
Anchorage, Alaska 99502-1896
Phone: (907) 243-2153

*ARIZONA, Douglas Office
Federal Communications Commission
P.O. Box 6
Douglas, Arizona 85608-0006
Phone: (602) 364-8414

CALIFORNIA, San Diego Office
Federal Communications Commission
Interstate Office Park
4542 Ruffner Street, Room 370
San Diego, California 92111-2216
Phone: (619) 467-0549

*CALIFORNIA, Livermore Office
Federal Communications Commission
P.O. Box 311
Livermore, California 94551-0311
Phone: (510) 447-3614

CALIFORNIA, Los Angeles Office
Federal Communications Commission
Cerritos Corporate Tower, Room 660
18000 Studebaker Road.
Cerritos, California 90701-3684
Phone: (310) 809-2096

CALIFORNIA, San Francisco Office
Federal Communications Commission
3777 Depot Road, Room 420
Hayward, California 94545-1914
Phone: (510) 732-9046

COLORADO, Denver Office
Federal Communications Commission
165 South Union Blvd., Suite 860
Lakewood, Colorado 80228-2213
Phone: (303) 969-6497/8

*FLORIDA, Vero Beach Office
Federal Communications Commission
P.O. Box 1730
Vero Beach, Florida 32961-1730
Phone: (407) 778-3755

FLORIDA, Miami Office
Federal Communications Commission
Rochester Building, Room 310
8390 N.W. 53rd Street
Miami, Florida 33166-4668
Phone: (305) 526-7420

FLORIDA, Tampa Office
Federal Communications Commission
2203 N. Lois Avenue, Room 1215
Tampa, Florida 33607-2356
Phone: (813) 228-2872

GEORGIA, Atlanta Office
Federal Communications Comission
3575 Koger Blvd
Koger Center-Gwinnett, Room 320
Duluth, Georgia 30136-4958
Phone: (404) 279-4621

*GEORGIA, Powder Springs Office
Federal Communications Commission
P.O. Box 85
Powder Springs, Georgia 30073-0085
Phone: (404) 943-5420

HAWAII, Honolulu Office
Federal Communications Commission
P.O. Box 1030
Waipahu, Hawaii 96797-1030
Phone: (808) 677-3318

ILLINOIS, Chicago Office
Federal Communications Commission
Park Ridge Office Center, Room 306
1550 Northwest Highway
Park Ridge, Illinois 60068-1460
Phone: (312) 353-0195

LOUISIANA, New Orleans Office
Federal Communications Commission
800 West Commerce Road, Room 505
New Orleans, Louisiana 70123-3333
Phone: (504) 589-2095

*MAINE, Belfast Office
Federal Communications Commission
P.O. Box 470
Belfast, Maine 04915-0470
Phone: (207) 338-4088

MARYLAND, Baltimore Office
Federal Communications Commission
1017 Federal Building
31 Hopkins Plaza
Baltimore, Maryland 21201-2802
Phone: (410) 962-2728

*MARYLAND, Laurel Office
Federal Communications Commission
P.O. Box 250
Columbia, Maryland 21045-9998
Phone: (301) 725-3474

MASSACHUSETTS, Boston Office
Federal Communications Commission
1 Batterymarch Park
Quincy, Massachusetts 02169-7495
Phone: (617) 770-4023

*MICHIGAN, Allegan Office
Federal Communications Commission
P.O. Box 89
Allegan, Michigan 49010-9437
Phone: (616) 673-2063

MICHIGAN, Detroit Office
Federal Communications Commission
24897 Hathaway Street
Farmington Hills, Michigan 48335-1552
Phone: (313) 226-6078

MINNESOTA, St. Paul Office
Federal Communications Commission
693 Federal Bldg. & U.S. Courthouse
316 North Robert Street
St. Paul, Minnesota 55101-1467
Phone: (612) 290-3819

MISSOURI, Kansas City Office
Federal Communications Commission
8800 East 63rd Street, Room 320
Kansas City, Missouri 64133-4895
Phone: (816) 353-3773

*NEBRASKA, Grand Island Office
Federal Communications Commission
P.O. Box 1588
Grand Island, Nebraska 68802-1588
Phone: (308) 381-4721

NEW YORK, Buffalo Office
Federal Communications Commission
1307 Federal Building
111 West Huron Street
Buffalo, New York 14202-2398
Phone: (716) 846-4511

NEW YORK, New York Office
Federal Communications Commission
201 Varick Street
New York, New York 10014-4870
Phone: (212) 620-3437/8

OREGON, Portland Office
Federal Communications Commission
1782 Federal Building
1220 S.W. Third Avenue
Portland, Oregon 97204-2898
Phone: (503) 326-4114/5

PENNSYLVANIA, Philadelphia Office
Federal Communications Commission
One Oxford Valley Office Bldg, Room 404
2300 East Lincoln Highway
Langhorne, Pennsylvania 19047-1859
Phone: (215) 752-1324

PUERTO RICO, San Juan Office
Federal Communications Commission
US Federal Building, Room 747
Hato Rey, Puerto Rico 00918-1731
Phone: (809) 766-5567

TEXAS, Dallas Office
Federal Communications Commission
9330 LBJ Freeway, Room 1170
Dallas, Texas 75243-3429
Phone: (214) 235-3369

TEXAS, Houston Office
Federal Communications Commission
1225 North Loop West, Room 900
Houston, Texas 77008-1775
Phone: (713) 861-6200

*TEXAS, Kingsville Office
Federal Communications Commission
P.O. Box 632
Kingsville, Texas 78364-0632
Phone: (512) 592-2531

VIRGINIA, Norfolk Office
Federal Communications Commission
1200 Communications Circle
Virginia Beach, Virginia 23455-3725
Phone: (804) 441-6472

*WASHINGTON, Ferndale Office
Federal Communications Commission
1330 Loomis Trail Road
Custer, Washington 98240-9303
Phone: (206) 354-4892

WASHINGTON, Seattle Office
Federal Communications Commission
11410 NE. 122nd Way, Suite 312
Kirkland, Washington 98034-6927
Phone: (206) 821-9037

*Licenses and examinations are not available at * locations

References

THE COMPLETE FCC HOME-STUDY COURSE
By Warren Weagant
COMMAND PRODUCTIONS, P.O. Box 2824, San Francisco, CA 94126-2824

A complete set of training lessons on audio cassette tape recordings and training manuals for the FCC General Radiotelephone Operator License. This self-study course provides beginners with fundamental concepts and comprehensive preparation for passing the federal FCC examination.

Many students have discovered that both reading training manuals and listening to special cassette recorded material makes learning faster, easier and more permanent.

Cassette tapes permit you to hear and absorb FCC material at any time and at any place you choose. Listen in office, home or car to turn unproductive time into FCC license training sessions!

You will find step-by-step solutions and answers for FCC Math problems too. Created specifically for those people who need a better understanding of vital basics, and then builds question by question in such a way that all solutions are understood - even by students with little or no electronics background.

FREE brochure and details about this course available directly from Command Productions. (See "Information Request Form" on page 223).

Reader Comments and Feedback

Dear Reader:

Your comments are always appreciate and help us to continue to publish the most up to date and relevant study materials.

Please let us know how this testing manual works for you. The response you send to us will help form the content of future editions of this study guide.

1. Where did you take the FCC examination? ________________

2. What date did you take the test? ______________________

3. What FCC test Elements did you take? ___________________

4. Your score: ____________

5. Did you find any material on the FCC exam that was not included in this testing manual? If so, what questions did you find difficult to answer?

   ____

   ____

   ____

6. Are you interested in obtaining any other FCC License? If so,

   which are you interested in obtaining? ____________________

7. Do you have any suggestions for improving these study materials more effective and easier to understand?

   ____

YOUR NAME: ____

ADDRESS: ____

CITY; ____________________ STATE: _____ ZIP: __________

Mail to: Warren Weagant, COMMAND PRODUCTIONS
P. O. Box 2824, San Francisco, CA. 94126-2824

Information Request Form

Mail to:

COMMAND PRODUCTIONS
FCC License Training
P.O. Box 2824
San Francisco, CA. 94126-2824

Please rush free ordering information for the FCC License Course materials checked below:

( ) The Complete FCC License Home-Study Course

( ) Tests-Answers for FCC General Radiotelehone Operator License (this testing manual).

Please send information to:

YOUR NAME: ______________________________

ADDRESS: ______________________________

CITY: ______________ STATE: ______ ZIP: ________

Check here ( ) if you also want information and quantity prices for a school or other educational institution.

Check here ( ) if you would like us to send information about this FCC testing manual to your friends or associates. Please attach their names and addresses.